Fuzzy Optimization Techniques in the Areas of Science and Management

This book helps to enhance the application of fuzzy logic optimization in the areas of science and engineering. It includes implementation and models and paradigms, such as path planning and routing design for different wireless networks, organization behavior strategies models, and so forth. It also:

- Explains inventory control management, uncertainties management, loss minimization, game optimization, data analysis and prediction, different decision-making system and management, and so forth
- Describes applicability of fuzzy optimization techniques in areas of science and management
- Resolves several issues based on uncertainty using member function
- Helps map different problems based on mathematical models
- Includes issues and problems based on linear and nonlinear optimizations
- Focuses on management science such as manpower management and inventory planning

This book is aimed at researchers and graduate students in signal processing, power systems, systems and industrial engineering, and computer networks.

Computational Intelligence in Engineering Problem Solving

Series Editor
Nilanjan Dey

Computational Intelligence (CI) can be framed as a heterogeneous domain that harmonized and coordinated several technologies, such as probabilistic reasoning, artificial life, multi-agent systems, neuro-computing, fuzzy systems, and evolutionary algorithms. Integrating several disciplines, such as Machine Learning (ML), Artificial Intelligence (AI), Decision Support Systems (DSS), and Database Management Systems (DBMS) increases the CI power and impact in several engineering applications. This book series provides a well-standing forum to discuss the characteristics of CI systems in engineering. It emphasizes on the development of CI techniques and their role as well as the state-of-the-art solutions in different real world engineering applications. The book series is proposed for researchers, academics, scientists, engineers and professionals who are involved in the new techniques of CI. CI techniques including artificial fuzzy logic and neural networks are presented for biomedical image processing, power systems, and reactor applications.

Applied Machine Learning for Smart Data Analysis
Nilanjan Dey, Sanjeev Wagh, Parikshit N. Mahalle, and Mohd. Shafi Pathan

IoT Security Paradigms and Applications
Research and Practices
Sudhir Kumar Sharma, Bharat Bhushan, and Narayan C. Debnath

Applied Intelligent Decision Making in Machine Learning
Himansu Das, Jitendra Kumar Rout, Suresh Chandra Moharana, and Nilanjan Dey

Machine Learning and IoT for Intelligent Systems and Smart Applications
Madhumathy P, M Vinoth Kumar, and R. Umamaheswari

Industrial Power Systems: Evolutionary Aspects
Amitava Sil and Saikat Maity

Fuzzy Optimization Techniques in the Areas of Science and Management
Edited by Santosh Kumar Das and Massimiliano Giacalone

Semantic Web Technologies
Edited by Archana Patel, Narayan C. Debnath, and Bharat Bhushan

For more information about this series, please visit: *www.routledge.com/ Computational-Intelligence-in-Engineering-Problem-Solving/book-series/CIEPS*

Fuzzy Optimization Techniques in the Areas of Science and Management

Edited by
Santosh Kumar Das and Massimiliano Giacalone

CRC Press
Taylor & Francis Group
Boca Raton   London

CRC Press is an imprint of the
Taylor & Francis Group, an **informa** business

First edition published 2023
by CRC Press
6000 Broken Sound Parkway NW, Suite 300, Boca Raton, FL 33487–2742

and by CRC Press
4 Park Square, Milton Park, Abingdon, Oxon, OX14 4RN

CRC Press is an imprint of Taylor & Francis Group, LLC

© 2023 selection and editorial matter, Santosh Kumar Das and Massimiliano
Giacalone; individual chapters, the contributors

Library of Congress Cataloging-in-Publication Data
A catalog record for this book has been requested

ISBN: 978-1-032-34286-3 (hbk)
ISBN: 978-1-032-34347-1 (pbk)
ISBN: 978-1-003-32171-2 (ebk)

DOI: 10.1201/b23138

Typeset in Times New Roman
by Apex CoVantage, LLC

Contents

About the Editors vii
List of Contributors ix
Preface xi

SECTION 1   Energy Resources Management

Chapter 1   Routing Recovery Protocol for Wireless Sensor Network
Based on PSO-ACO and Neural Network 3

Jeevan Kumar, Tapan Kumar Dey, Rajesh Kumar Tiwari,
and Amit Kumar Singh

Chapter 2   Fuzzy Inference-Based Optimal Route-Selection Technique
in Wireless Ad Hoc Network 23

Santosh Kumar Das, Nikhil Patra, Biswa Ranjan Das,
and Aditya Sharma

Chapter 3   Fuzzy-Based Mathematical Model for Optimizing Network
Lifetime in MANET 41

Manoj Kumar Mandal, Arun Prasad Burnwal,
B. K. Mahatha, and Abhishek Kumar

SECTION 2   Modelling and Aggregation

Chapter 4   Game Theory–Based Conflicting Strategy Management
Technique in Wireless Sensor Network 57

Santosh Kumar Das, Aman Kumar Tiwari, Somnath Rath,
and Joydev Ghosh

Chapter 5   Cluster-Based Routing Protocol for WSN Using Fusion of
Swarm Intelligence and Neural Network 75

Jeevan Kumar, Rajesh Kumar Tiwari, Tapan Kumar Dey,
and Amit Kumar Singh

Chapter 6   Nonlinear Fuzzy Optimization Technique for WSN
Based on Quadratic Programming 93

Manoj Kumar Mandal, Arun Prasad Burnwal,
B. K. Mahatha, and Abhishek Kumar

SECTION 3   Data Analysis and Prediction

Chapter 7   A Review Based on Prediction Analysis to Mitigate
the Issues of COVID-19 117

Santosh Kumar Das and Aditya Sharma

Chapter 8   Data Analysis and Prediction for WSN Based on Linear
and Quadratic Optimization Techniques 137

Manoj Kumar Mandal, Arun Prasad Burnwal,
B. K. Mahatha, and Abhishek Kumar

Chapter 9   Machine Learning-Based Data Analysis for Managing
Challenges of COVID-19: A Survey 155

Santosh Kumar Das and Joydev Ghosh

Chapter 10 Fuzzy Geometric-Based Cost-Optimization Technique
for Company 177

Neha Kumari, Arun Prasad Burnwal, and Neha Keshri

Index 195

About the Editors

Santosh Kumar Das received his PhD degree in computer science and engineering from Indian Institute of Technology (ISM), Dhanbad, India, in 2018 and completed his MTech degree in computer science and engineering from Maulana Abul Kalam Azad University of Technology (erstwhile WBUT), West Bengal, India, in 2013. He has about three years' teaching experience as Assistant Professor at the School of Computer Science and Engineering, National Institute of Science and Technology (Autonomous), Institute Park, Pallur Hills, Berhampur, Odisha, India. He is currently working as Assistant Professor at the Department of Computer Science and Engineering, Sarala Birla University, Birla Knowledge City, PO-Mahilong, Purulia Road, Ranchi, India. He has more than eight years' teaching experience. He has authored/edited five books with Springer, including the series *Lecture Notes in Networks and Systems, Tracts in Nature-Inspired Computing*, and *Studies in Computational Intelligence.* He has contributed more than thirty-five research papers. His research interests mainly focus on ad hoc and sensor network, artificial intelligence, soft computing, and mathematical modeling. His h-index is 16, with more than seven hundred citations.

Google Scholar Profile: https://scholar.google.com/citations?user=AkQx5Ko AAAAJ&hl=en.

Massimiliano Giacalone received his PhD degree in computational statistics and applications from the University of Naples Federico II, Department of Mathematics and Statistics. He received his graduate degree in statistics and economics sciences, magna cum laude, from the Faculty of Economics, University of Palermo. He is currently a researcher in statistics and teaching staff member of the Department of Economics and Statistics, University of Naples Federico II. He has contributed more than one hundred research papers. His research area encompasses the following subjects: multidimensional data analysis, big data for social statistics, Norm-p linear and nonlinear regression, permutation tests, control charts and economic statistics, and applications of statistics in medicine, justice, and finance.

Google Scholar Profile: https://scholar.google.com/citations?hl=en&user=O9 12OOMAAAAJ&view_op=list_works&sortby=pubdate

Contributors

Arun Prasad Burnwal
GGSESTC
Bokaro, Jharkhand, India

Biswa Ranjan Das
National Institute of Science and
    Technology (Autonomous),
Berhampur, Odisha, India

Santosh Kumar Das
Sarala Birla University
Ranchi, Jharkhand, India

Tapan Kumar Dey
R.V.S College of Engineering &
    Technology
Jamshedpur, India

Joydev Ghosh
National Research Tomsk Polytechnic
    University (TPU)
Tomsk, Russia

Neha Keshri
Jharkhand Rai University
Ranchi, India

Abhishek Kumar
NIT
Jamshedpur, Jharkhand, India

Jeevan Kumar
R.V.S College of Engineering &
    Technology
Jamshedpur, India

Neha Kumari
Jharkhand Rai University
Ranchi, India

B. K. Mahatha
Amity university
Jharkhand, Ranchi, India

Manoj Kumar Mandal
Jharkhand Rai University,
Ranchi, India

Nikhil Patra
National Institute of Science and
    Technology (Autonomous),
Berhampur, Odisha, India

Somnath Rath
National Institute of Science and
    Technology (Autonomous),
Berhampur, Odisha, India

Aditya Sharma
Institute of Nanoengineering and
    Microsystems
National Tsing Hua University
Hsinchu, Taiwan

Amit Kumar Singh
SRM University AP
Guntur District, Andhara Pradesh, India

Aman Kumar Tiwari
National Institute of Science and
    Technology (Autonomous),
Berhampur, Odisha, India

Rajesh Kumar Tiwari
R.V.S College of Engineering &
    Technology
Jamshedpur, India

Preface

In the modern era, most applications are based on imprecise information along with nonlinear parameters. It creates several types of issues based on uncertainties. The main reason for these issues are the different types of conflict in the choices and strategies of the customers. Several problems are solved and managed with the context of a rigid point of view. But it is not sufficient to deal with real-life applications, because real-life applications change frequently based on the demand of the society. Fuzzy logic is an intelligent technique under the guidance of soft computing as well as artificial intelligence. It works with several types of membership functions that help map data into some intervals. It helps to reduce imprecise information efficiently and helps to control uncertainties. The application of fuzzy logic has increased rapidly based on the requirement of the customers in different areas. Hence, the proposed book illustrates several applications of fuzzy logic in the areas of science as well as in engineering.

THE OBJECTIVE OF THE BOOK

The main objective of this book is to enhance the application of fuzzy logic optimization in the areas of science and engineering. It helps to resolve some problems based on methodologies of fuzzy optimization. It generates feasible as well as optimal solution based on the proposed problem. The content of this book helps the reader in the areas of academics as well as research to enhance the models of real-life solution.

ORGANIZATION OF THE BOOK

The book contains ten chapters that are organized into three sections, as follows: Section 1 discusses some applications of fuzzy optimization based on energy resource management. Section 2 discusses models and aggregation of data based on fuzzy logic. Section 3 discusses several types of data analysis and prediction based on fuzzy logic optimization.

SECTION 1: ENERGY RESOURCES MANAGEMENT (CHAPTERS 1–3)

This section helps in analyzing and discussing fuzzy optimization based on the energy resource management system.

CHAPTER 1

This chapter discusses a routing technique based on the fusion of two nature-inspired optimization techniques, ACO and PSO. The combination of both helps optimize the network by using NN. It indicates neural network that is based on an artificial intelligence technique. It helps optimize network metrics and provides the optimal route.

CHAPTER 2

This chapter discusses models of fuzzy logic with an inference system. This fuzzy logic helps in modeling several issues and inference efficiently with the membership function. It also helps in modeling the application based on an optimal and feasible path by using fuzzy logic in the network.

CHAPTER 3

This chapter is based on network lifetime enhancement of mobile ad hoc network using fuzzy logic. In this paper, fuzzy logic works with several types of membership functions along with some variables. This variable helps in modeling applications based on linguistic nature. It optimizes both linear and nonlinear parameters based on the objective function of the network.

SECTION 2: MODELING AND AGGREGATION (CHAPTERS 4–6)

This section helps in modeling several applications based on some applications of aggregation methods.

CHAPTER 4

This paper is based on strategy optimization of sensor network by the fusion of game theory and fuzzy logic. Game theory acts as an intelligent technique for managing and controlling different conflicting strategies of the network efficiently. Fuzzy logic helps in producing membership functions and controlling the imprecise nature of the network.

CHAPTER 5

This paper is based on cluster analysis and its models for sensor network. The work is based on some fusion of swarm intelligence. This intelligence is mixed with neural network for modeling cluster analysis. It helps in efficiently communicating with different nodes based on the service of the sensor network.

CHAPTER 6

This chapter is based on fuzzy optimization of sensor network for analysis and design of network modeling. It helps in designing several applications of the network based on the fusion of fuzzy logic and quadratic programming. It helps in modeling several objective functions based on constraint analysis.

SECTION 3: DATA ANALYSIS AND PREDICTION (CHAPTERS 7–10)

This section helps in designing some applications based on fuzzy analysis with predictions. This prediction helps in modeling some analysis based on data and its related services.

CHAPTER 7

This chapter is used for review analysis based on prediction for COVID analysis. This review is prepared for solving several issues along with different types of diseases. It helps in modeling several information for disease analysis. It incorporates several intelligent techniques that provide intelligent outcomes based on services.

CHAPTER 8

This chapter is based on the optimization technique of sensor network for the analysis of different types of data. It helps in modeling several applications based on the services. This modeling is based on the fusion of linear and quadratic programming that helps map efficient service into the network.

CHAPTER 9

This chapter is based on several types of data analysis based on some managing information based on COVID-19. It helps in providing a detailed information based on challenging information with the context of the disease. It helps in predicting and managing several challenges based on data analysis and its services.

CHAPTER 10

This chapter is based on the cost optimization of the company. It deals with several optimization strategies for modeling applications. It helps in dealing with several types of analysis based on the objective function and its related constraint. The constraint helps in controlling the strategy of the company based on modeling information.

Section 1

Energy Resources Management

1 Routing Recovery Protocol for Wireless Sensor Network Based on PSO-ACO and Neural Network

Jeevan Kumar, Tapan Kumar Dey, Rajesh Kumar Tiwari, and Amit Kumar Singh

CONTENTS

1.1 Introduction 4
1.2 Literature Review 4
1.3 Proposed Method 9
    1.3.1 Particle Swarm Optimization Protocol 9
    1.3.2 Ant Colony Optimization (ACO) Protocol 10
        1.3.2.1 Calculation of Pheromone Density 10
    1.3.3 Proposed Scheme PSO-ACO and NN 10
        1.3.3.1 REQ_ANT 11
        1.3.3.2 ACK 11
        1.3.3.3 Routing Table 11
    1.3.4 Working Principle 13
        1.3.4.1 Scenario: No Link and Node are Failure 14
        1.3.4.2 Second Scenario: Link is Failure 15
        1.3.4.3 Third Scenario: Node is Failure 15
    1.3.5 Work Flow Design 15
    1.3.6 Route Estimation by Fuzzy Logic 17
1.4 Experimental Evaluation and Analysis 18
    1.4.1 Simulation Environment 18
    1.4.2 Construct a Wireless Sensor Network System 18
1.5 Conclusion 20
1.6 References 20

DOI: 10.1201/b23138-2

1.1   INTRODUCTION

Wireless sensor network (WSN) is a very popular network that works as a wireless network [1, 2]. It is used to communicate each and others based on several purposes and activities. There are several components of WSN which are used to deal with several other components and parameters, such as base station (or BS), central point, source node, sink node, destination node, etc. Each component is crucial in the formation of a network communication that helps in modeling several applications. Several works have been proposed to explain different problems and the modeling of WSN [3, 4]. The current proposal is based on a route recovery system and modeling that helps in modeling several applications [5]. The communication should be adequate and efficient, because there are several types of intrusion and imprecise information in the system for modeling for analysis.

The contribution of the paper is based on motivation analysis and a design that helps in modeling several purposes and analyses. It uses several nature-inspired optimization algorithms, such as neural network, particle swarm optimization, and ant colony optimization, for the purpose of modeling. This modeling is based on an efficient design of network and analysis for an efficient route design. This route design helps in modeling several applications based on estimation analysis of a system that helps in route recovery system. This route recovery system helps in modeling several parameters, including the ant colony optimization as well as the particle swarm optimization systems, based on neural network systems. The combination of these inherent nature-inspired optimization systems is used to deal with several network lifetime analysis. It helps in modeling several route recovery and modeling systems to help design an efficient route.

Sections in the rest of the paper are based on some disciplines. The first section is based on the description of existing works that are used in WSN. The next section is based on the proposed method, which is the fusion of several nature-inspired algorithms. The next section is based on experimental analysis based on evaluation purposes. The last section is used for the conclusion, which concludes the paper.

1.2   LITERATURE REVIEW

In the last few years, several works have been proposed for wireless sensor network optimization and enhancement dealing with several parameters and optimization. In this section, some of the works are designed for enhanced network based on different information. Also, some of the works are discussed within the context of nature-inspired optimization technique system and its modeling. Pahar et al. [6] proposed a method for cough analysis and its different types of classification based on COVID-19. The work is based on machine learning algorithm and its different inherent elements. The work also used some of smartphone systems for resolving the issues. The disease is based on two types: first is normal cough, and the second is forced cough. It is based on some dataset used for COVID-19 patients. This article also illustrates that manpower or workload is to be optimized because day by day, in that situation, the numbers of patients increases simultaneously. There are several techniques used for machine learning, such as multilayer perceptron, support vector machine, convolution neural network, k-nearest neighbors algorithm. Finally, in this work, two types of coughs are analyzed; one is based on positive case, and the second is on normal case. Parvin and Vasanthanayaki [7] proposed an optimization-based technique for efficient routing in WSN. The purpose of this proposed method is to track the behavior of animals. This work is also based on a distributed system of the network. The key element of this paper is entropy technique, which is part of multicriteria decision-making. Finally, clustering is evaluated with the help of maximum finding mechanism of the entropy. Singh et al. [8] proposed an adaptive method featuring an energy-aware system that is used for communication purposes. This method is used for a physical communication system that is based on a delay-tolerant system. It helps in modeling application based on a cyber-physical system that is used for wireless sensor network. It helps in modeling application based on a delay-tolerant and prediction system. The application is based on mobile application for several operations. The work is simulated based on a network environment for predicting network lifetime. De et al. [9] proposed an illustration that tackles application management services and their challenges. This service is based on the application of a wireless sensor network system and its variations. It helps in modeling several challenges and application management. It helps in dealing with several algorithms of wireless network based on variation and its analysis. It helps in guiding several working principles and information system based on service management. It helps in dealing with algorithm analysis that helps manage several applications of the system. Zhu et al. [10] designed a technique by fusing game theory and Markov chain for improving network lifetime of the WSN. The combination of both techniques helps control and manage several multimedia elements, such as image, video stream, voice, etc. This Markov model is based on a hidden Markov model that is used to enhance the accuracy of the game in the network, which means increase in the transmission of data packet. Gulati et al. [11] illustrated a method based on tweet analysis on different information related to COVID-19 and its different pandemic effects. The method is based on several types of analysis and information that deals with machine learning technique. The work is based on a comparative analysis which is done on sentiment analysis technique, which is based on machine learning. The dataset used in this model is first analyzed based on a traditional method known as lexicon-based approach, then it is extended with the help of sentiment analysis. Das et al. [12] designed a book to show an architectural solution system based on wireless network. This service is not only based on network; it is also based on several systems and information based on the architecture of the network. It helps in modification based on the architecture of the system. It helps model several issues of the network. Several issues are used and deal with the system. Some issues are mentioned, such as energy efficiency system, network lifetime system, resource management, data aggregation system, etc. It helps to model several solutions and the security management of wireless network. Singh et al. [13] designed an efficient modeling method that helps vehicular communication. The work is based on communication purpose and modeling. It helps in several applications that helps to model several vehicular relay management systems for an efficient network communication. The work is based on a utilization purpose that helps in modeling several potential system managements. It helps in modeling several relay applications and network modeling. It helps in predicting several vehicular communications based on selfish note prediction. It helps in strategy management for tracking network performance. Esposito and Choi [14] designed a strategy-based technique for WSN. This is based on localization system of the network. It helps in alerting the nodes as secure location and nonsecure location with the help of the anchor system, which indicates here the known position of the nodes. It also helps in reducing the cost of the network by decreasing the exhaust system of the network. De Fátima Cobre [15] proposed a method based on severity analysis of COVID-19. In this article, several types of predictions and diagnoses are done based on several information. The work is based on biochemical test analysis of a machine learning algorithm. It helps in analyzing different information of the stated issue based on machine learning. The work is a fusion of several types of algorithms, such as artificial neural network, k-nearest neighbor algorithm, and decision tree, to improve the results. Das et al. [16] proposed a book that helps in modeling several applications based on industrial application and formulation. It helps in modeling several information based on machine learning systems and modeling. The content of this book deals with several information based on the decision-making system and the prediction analysis. It helps in dealing with some applications based on natural language processing, machine learning, computer vision, image processing, etc. Each application and system is based on several information and modeling systems for dealing with and the analysis of information modeling. Singh and Pamula [17] designed a method showing an intelligent communication and routing modeling. The work is based on the vehicular communication system based on strategy management. It helps, based on strategy, in behavior modeling and analysis to help enhance network lifetime. The application is deployed on several purposes related to protocol management. It helps in tracking analysis in novel behavior based on delay-tolerant management. It helps to utilize and analyze the system based on vehicular communication. It helps in outperforming the result based on several variations. Phoemphon et al. [18] designed a hybrid method for optimization in WSN using the fusion of PSO, fuzzy logic, and machine learning. This fusion helps in finding global positioning systems of the network. The main key issue of this system is residual energy of the sensor node, and the fusion helps to reduce it comparatively for the network. This fusion also helps in analyzing several factors of the network, such as variation of the topology, coverage sensing, and density of the nodes. Kassania et al. [19] designed a method based on COVID-19 which is part of the automatic detection of the COVID disease. Two basic components are used in this article, such as CT image and X-ray; based on these two medical information, diseases are detected. A machine learning tool is used to detect the information. There are several symptoms used in this model, such as fever, sore throat, and cough. Additionally, there are several machine learning algorithms used in this article, such as ResNet, DenseNet, MobileNet, NASNet, etc., for predicting the disease efficiently and effectively. The optimization techniques used in COVID-19 management help in modeling several wireless network optimization and nature-inspired optimization techniques. De et al. [20] designed a book explaining wireless sensor network. It helps model several applications based on services and management. It helps deal with several information and management systems for key areas of management. The work is based on nature-inspired application and computing that helps model several issues. It helps in the implementation of several applications and computation information systems. Information in this book is distributed in the form of bio- and nature-inspired systems. It helps model and design several applications for single-objective and multiobjective optimization systems. Singh and Pamula [21] designed a delay management for a communications-based system. The work is based on machine learning-based application that helps in several purposes. The work is based on several classifier model of analysis based on solution for vehicular network. It helps deal with several network metrics based on certain parameters of network. The work is based on strategy management for handling the inadequate design of the system. It helps outperform the result based on the consideration of the system modeling. Wei et al. [22] designed a PSO-based localization method for WSN. In this work, network topology is based on three-dimensional method. This dimensional method helps enhance the network topology efficiently. It efficiently finds the distance between two nodes, whether these nodes are source, destination, anchor, unknown, or a simple node. Finally, it helps enhance the network metrics efficiently in terms of some existing works. Tamal et al. [23] proposed a method called rapid early system based on the diagnosis of different types of X-ray. This X-ray is based on chest analysis and predictions based on different radiomics systems. The work is based on a framework system based on different integration systems. The work is totally based on the COVID-19 disease from different dependent datasets. The images of this analysis are based on pneumonia and normal lung. Z-scoring and the bagging model tree are used for the analysis and management. Das et al. [24] designed a system and model for wireless network and application. It helps in modeling several information based on service management and application modeling. The work is based on several information and management systems, such as energy resource management and modeling. It helps in dealing with several security and privacy management that helps design and its enhancement. The work is based on troubleshooting and automation system for network lifetime management and its enhancement. It gives several protocols for design perspective and modeling. Singh et al. [25] designed a method of strategy management that is based on delay analysis. It helps in maintaining several network managements based on some replication analysis. It helps in dealing with and managing several packet control management systems. It helps in managing several data delivery and resource management for communication systems handling. The work is based on routing analysis that helps deal with and manage intermediate node analysis systems. The work deals with several assists for replication management. This replication is based on the control system and the parameter dealing system. Lu et al. [26] designed a PSO-based technique for aggregation of data in WSN. In this work, PSO is used to generate a tree, known as a spanning tree. This tree generation helps reduce two things, such as energy consumption and redundancy of data packets. This is also used in the Pareto optimal technique of PSO for improving the results of the network using multiobjective optimization, such as interference in communication, timing period of the communication, and network lifetime. Lam et al. [27] designed a model for precision analysis of a model based on medicine analysis and its pre-scribing system. The model is based on COVID-19 and its different factors of diseases. The information is also collected from pharmacotherapy management based on different factors. In this method, gradient boost tree is used as machine learning model based on several factors and parameters. The algorithm is based on parameters of gradient boost tree that are based on two basic analyses. First is indicator, which is based on patients that are in need of and survive with oxygen. Second are patients that survive based on treatment. Das et al. [28] designed an application management illustration for the purpose of wireless network systems and services. It helps in modeling several applications based on different security and challenge system management. It helps in issues of management based on several parts of the communication system. It helps in adopting several solutions based on some higher analysis and management. The work is based on complexity management, which helps in modeling several solutions with the context of management based on different variations of the wireless network. Tang et al. [29] proposed an optimization technique for WSN that uses PSO for video sensing in the network. This paper uses three basic elements, such as power of the nodes (i.e., residual energy), data rate, and network lifetime. The PSO technique uses the stochastic method for optimizing nonlinear parameters of the network. Finally, this method helps to optimize the overall network metrics by using suboptimal and dividing the main problem into some subproblems. Weng et al. [30] proposed a method for volatility forecasting system based on an online learning system and model. It is based on different factors and information based on regularization system. The work is based on the COVID-19 pandemic and its different variation systems and modeling. The method is based on genetic algorithm based on several learning methods. It helps in analyzing updated learning ability system based on some factors and information. S. K. Das [31] proposed a method for the purpose of application design system and its management. It helps in several application man-agement that helps to deal with several challenges and issue management. It helps in modeling several information systems based on smart application systems. It helps in modeling several security management systems based on emergency management and application systems. It helps in giving a new guideline that helps in modeling several issues in terms of solution. This solution helps to model several services based on real-life application management. It helps in several monitoring and application management systems based on an emergency and security modeling system based on services. Salehian and Subraminiam [32] designed a clustering-based PSO routing for WSN. This is based on unequal clustering system. The main objective of this algorithm is to optimize alive nodes of the WSN. In this method, a number of alive nodes increases gradually based on PSO optimization parameters. Finally, it ignores hot spot of the network and increases network parameters simultaneously. Huang et al. [33] proposed a data-driven method for COVID-19. The work is done in Pakistan at Lahore. The method is based on a machine learning technique based on different data-driven systems. The work is based on weakness analysis of different variations. The weakness is analyzed based on current test methodology. In this method, priority ranking and logistic regression are used for analysis purposes and for modeling it. The local data is analyzed based on the concept of data-driven method for optimal solution. Das et al. [34] designed a book for the purpose of smart application design. The work of this book is based on smart application with the fusion of smart computing. It helps in modeling several applications based on some service and management. It helps in modeling several nature-inspired applications based on computing application services. It helps in modeling and giving new sight in the subarea of network modeling, data analysis and prediction, network lifetime management, resource and energy management, etc. It helps in modeling several information systems for the management of dynamic application and planning and services.

1.3 PROPOSED METHOD

This is the main section of the proposed method that is based on the fusion of some nature-inspired optimization techniques and its related inherent element. This section is divided into some subsections that help to model the proposed goal efficiently.

1.3.1 PARTICLE SWARM OPTIMIZATION PROTOCOL

Particle swarm optimization is a well-known bioinspired technique that is used to address a variety of optimization problems in a variety of fields, including soft computing, artificial intelligence, data mining, robotics, wireless sensor networks, and many more. These are based on several types of optimizations and learning methodologies that are used on several areas [35–37]. The PSO determines the optimal solution for a particle's position and velocity while also minimizing a variety of other characteristics, such as path distance, communication cost, power consumption, end-to-end delay, and so on. Equations 1 and 2 are used to calculate the node's velocity and location, and they are utilized to update the node's position [38].

$$V_{id} = W \times V_{id} + C_1 R_1 (P_{id} - X_{id}) + C_2 R_2 (P_{gd} - X_{id}) \tag{1}$$

$$X_{id+1} = X_{id+1} + V_{id} \tag{2}$$

At the time T, the velocity and position of a particle is updated by equations 3 and 4.

$$V_{id(t+1)} = W \times V_{id(t)} + C_1 R_1 (P_{id} - X_{id(t)}) + C_2 R_2 (P_{gd} - X_{id(t)}) \tag{3}$$

$$X_{id(t+1)} = X_{(id+1)(t)} + V_{id(t)} \tag{4}$$

V_{id} is the velocity of the particle. X_{id} is the current position of the particle. W is the inertia weight. C_1 is the cognitive acceleration coefficient. C_2 is the social acceleration coefficient. P_{id} is the own best position of the particle. P_{gd} is the global best position among the group of particles. R_1 and R_2 are the uniformly distributed random numbers in the range (0 to 1). X_{id+1} is the modified position.

1.3.2 ANT COLONY OPTIMIZATION (ACO) PROTOCOL

The ACO protocol is based on pheromone density. The ACO algorithm uses mobile agents as ants to identify the most feasible and best path in a network. When ants construct paths from the beginning node to the destination node, they generate pheromones and deposit them on the traversed nodes.

1.3.2.1 Calculation of Pheromone Density

During the selection of the next hop for forwarding the route request to the destination, the nearest node is not always preferable. Let there exist a link from node i to node j. The pheromone density of the link Pd_{ij} can be calculated as equation 5:

$$Pd_{ij} \frac{\left(\text{Distance between the node } i \text{ to node } j\right)}{\left(\text{Time taken to reach node } i \text{ to node } j\right)} \tag{5}$$

1.3.3 PROPOSED SCHEME PSO-ACO AND NN

We present a reliable routing technique to find the shortest path for data transmission using parameters like neighbor node distance and its energy. The back propagation neural network algorithm is used to select the best nodes in a path of data transmission. Once the route is established, the transmission of data is processed. The back propagation neural data routing is combined with smart sampling to achieve an energy-efficient, reliable packet routing. The range or position estimation of the nodes is accomplished based on the communicating source and the destination nodes from small overhead signals collected at the base station from different sensor nodes. The route following the shortest path is computed for data transmission using the Euclidean distance formula, as in equation 6. In this section, some of the information are shown for the purpose of communication system of nodes based on route establishment. It contains the request, acknowledgement, routing table management, and its related algorithm.

$$D(s,d) = \sqrt{(x_d - x_s)^2 + (y_d - y_s)^2} \tag{6}$$

Equation 7 is used to compute the forwarding node's energy conditions.

$$E_r(t) = E_i(t) - (0.1 * D) \tag{7}$$

After this, we use PSO-ACO scheme for further processing of the data. Unlike the existing metrics, the proposed scheme considers a greater number of critical factors for optimal route selection. PSO-ACO constructs an optimal route from source to destination by considering the position of the node, the pheromone density, and the number of hops along the path. Each node keeps some information in its routing table. We are considering different scenarios for finding the optimal routing path, as shown in Figure 1.1. In this case, there is no link and no nodes are a failure. Link is a failure in Figure 1.2, and finally, in Figure 1.3, the node is a failure.

1.3.3.1 REQ_ANT
The source node sends the REQ_ANT by setting the value of pheromone density as 0. During the propagation of REQ_ANT over the network, every intermediate node adds its pheromone density with the REQ_ANT. Upon reaching the destination node, the packet's pheromone density field determines the level of connectivity of that route.

1.3.3.2 ACK
The destination node updates the pheromone density of ACK with the highest pheromone count value among multiple REQ_ANTs it received. Then it sends back ACK through the path having the highest pheromone density.

1.3.3.3 Routing Table
In the proposed scheme used, three fields have been added, namely position, pheromone count, and HCM. The pheromone count field stores the value of the pheromone count of the path from the source node to the current node. There are several

FIGURE 1.1   No link and node failure.

FIGURE 1.2   Link failure.

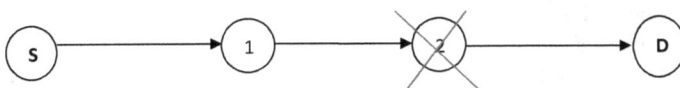

FIGURE 1.3   Node failure.

algorithms mentioned in this section for the purpose of efficient communication based on the proposed method and its formulations. Source node, intermediate node, destination node, link failure, and node failure are shown in Algorithms 1 to 5.

Algorithm 1: Source Node Communication

```
data send (s,d)      // s is the source node, d is the destination node
{
if (rt != 0)         // rt = 0 indicates no entry for destination d in routing table.
send data to d;
else
send_REQ_ANT (d);
}
/* source sends REQ_ANT */
send_REQ_ANT (d)
        {
        Find node_ position;
        calculate_Pd;
        updater_table;
        broadcast (p);
        }
```

Algorithm 2: Intermediate Node Communication

```
receive_REQ_ANT (packet p)
{
if (p.rq_dst = = intermediate node)
receive REQ_ANT;
create ACK and send to source node;
hopcount = p.hopcount +1;
calculate Pd;
update r_table;
send_REQ_ANT (d);
}
```

Algorithm 3: Destination Node Communication

```
dest_receive_REQ_ANT (packet p)
{
if (p.rq_dst = = destination node)
        {
        receive REQ_ANT;
        create ACK and send to sender node;
        update Pd;
        update r_table;
        }
}
```

Algorithm 4: Link Failure Method

Send_REQ_ANT (packet p)
{
If (node= = link exist)
{
Send_REQ_ANT (packet p)
}
Else
{
find position of new node which is nearest to destination using PSO equation.
update Pd;
update r_table;
}

Algorithm 5: Node Failure Method

Send_REQ_ANT (packet p)
{
If (node= = next node exist)
{
Send_REQ_ANT (packet p)
}
Else
{
find position of new node which is nearest to destination using PSO equation.
update Pd;
update r_table;
}

1.3.4 WORKING PRINCIPLE

When a source node wishes to transfer data to a destination, it first determines whether a path exists from source to destination. Data is transmitted through the path if one exists. The route-finding procedure will be started if this is not the case. The source node analyzes if its neighbor nodes are closest to the destination node based on node position, pheromone density, and hop count in this process. If that's the case, send REQ ANT data via that path. According to algorithm 1, the neighbors' routing table entries will be updated. When the REQ ANT data is received by the neighbor node, the ACK is then created and sent to the source node. The value of the REQ ANT's hop count field will be increased by 1. It's saved in the hop count field associated with that routing table entry. The pheromone density is then calculated. Algorithm 2 shows the method for the intermediate node forwarding procedure. The routing table is going to be updated. Create and send ACK to sender node when REQ ANT is received by destination node. Calculate the hop count and pheromone density, then update the routing table. Algorithm 3 shows the algorithm for the destination node procedure. The routing table is going to be updated. Figure 1.4 depicts an example routing and data transmission situation. The failure shown in algorithm 4 is the connection algorithm. If the link fails, use equations 1 and 2 to determine the position of the new node. Using pheromone density and hop count, determine the best node. Refresh the routing table. Figure 1.5 depicts an example scenario for connection failure routing.

The algorithm for when the node is failure is shown in Figure 1.6. If node is failure, then find the position of the new node according to equations 1 and 2. Find out the optimal node with the help of pheromone density and hop count. Update routing table.

Different scenarios for routing and data transfer are shown in following sub-section.

1.3.4.1 Scenario: No Link and Node are Failure

Assume that source node S is the source and destination node D is the destination. Let's pretend that node S wishes to deliver data to destination D. Node S determines whether a path to the destination exists. Let's pretend there's no way to go to D. Nodes 1, 3, and 5 are S's neighbors. Node S searches its neighbor table for entries from nodes 1, 3, and 5 and calculates their distance from the target. It also obtains the values of their corresponding additional parameters. These three neighbors all meet the required value. Then S calculates the equivalent phero-mone density (Pd) for nodes 1,3, and 5 in the routing table of node S, which is set to 0.

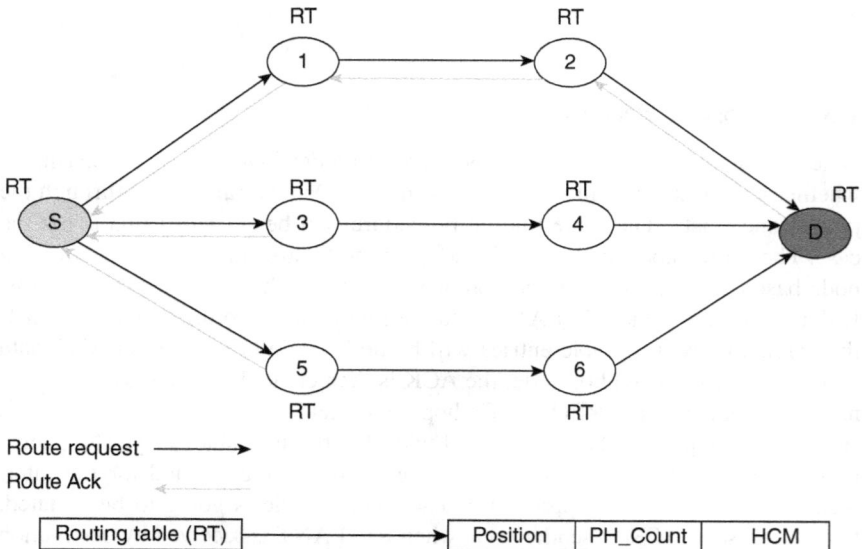

FIGURE 1.4 Scenario of route selection process of PSO-ACO in normal case.

Node S now transmits a REQ-ANT packet to the node closest to the target. When node 1 receives the REQ-ANT, it creates an ACK link to the source nodes S and updates the routing table. This process will continue till data reaches destination node D. Between the source S and the destination D, a path is constructed. S can now transfer data to D using the path (S-1-2-D). As illustrated in Figure 1.4, the suggested scheme takes into account node position, hop count, and pheromone density while determining the best route from source to destination.

1.3.4.2   Second Scenario: Link is Failure

Node 2 receives data in the event of a link failure, as seen in Figure 1.5. After this link fails, make a new link to a node that is the closest neighbor to node 2 and update the routing table. Between the source node S and the destination node D, a path is constructed. S can now transfer data to D using the path (S-1–2–4-D).

1.3.4.3   Third Scenario: Node is Failure

Figure 1.6 depicts a node failure condition, with node 2 failing. Return to the previous node and make a new link to the node that is closest to node 1, then update the routing table. Between the source node S and the destination node D, a path is constructed. S can now deliver data to D using the path (S-1–3–4-D).

1.3.5   WORK FLOW DESIGN

Figure 1.7 depicts the flow chart of the proposed technique for PSO-ACO and NN. When a source node wishes to transfer data to a destination, it first determines whether a path exists from source to destination. Data is transmitted through a path if one exists. Otherwise, the route discovery process will start. The source node analyzes if its neighbor nodes are closest to the destination node based on node position, pheromone density, and hop count in this process. If that's the

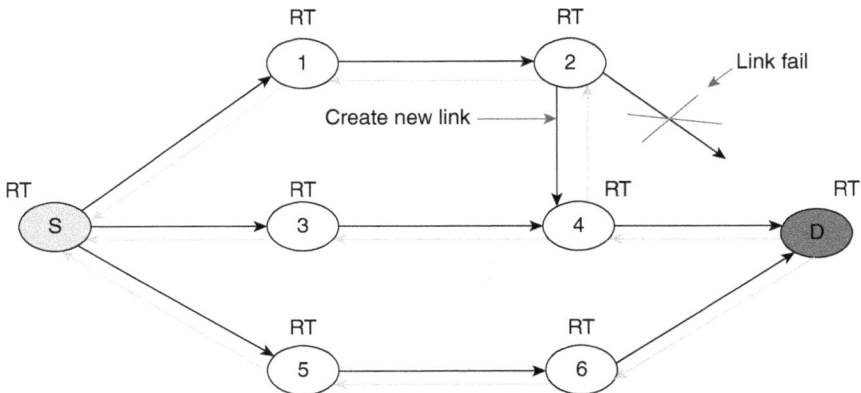

FIGURE 1.5   Scenario of route selection process of PSO-ACO in link failure.

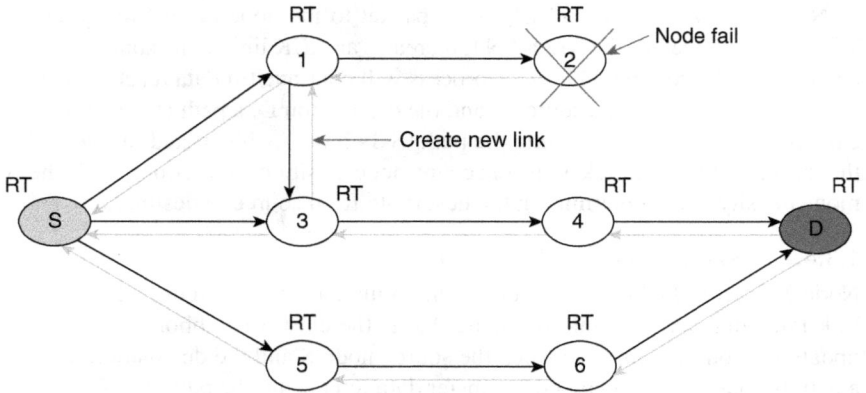

FIGURE 1.6   Scenario of route selection process of PSO-ACO in node failure.

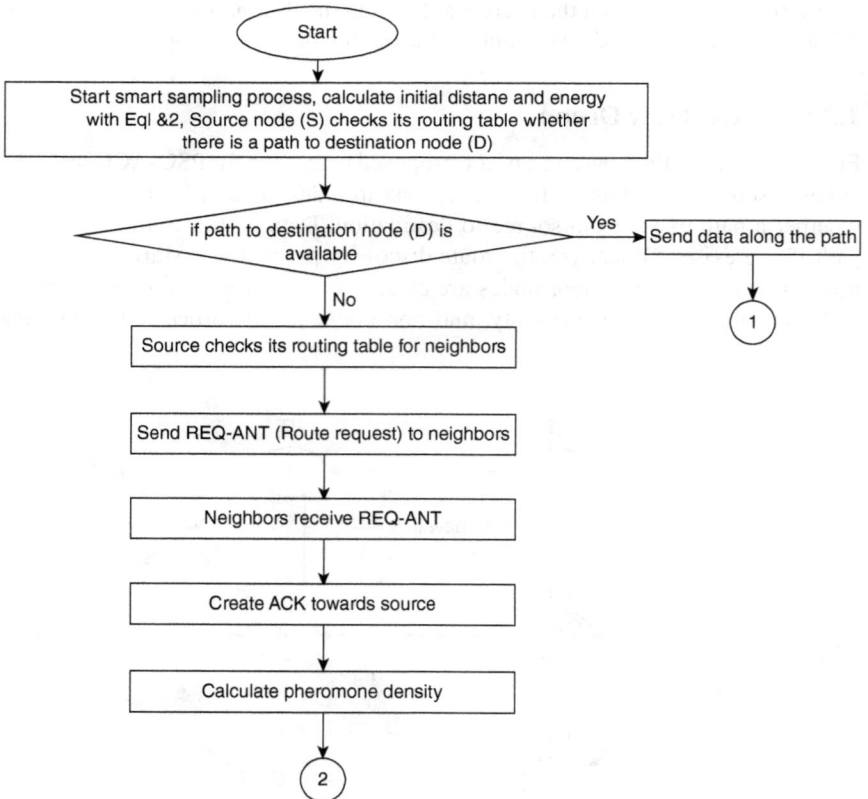

FIGURE 1.7   Flow chart of routing process through PSO-ACO and NN.

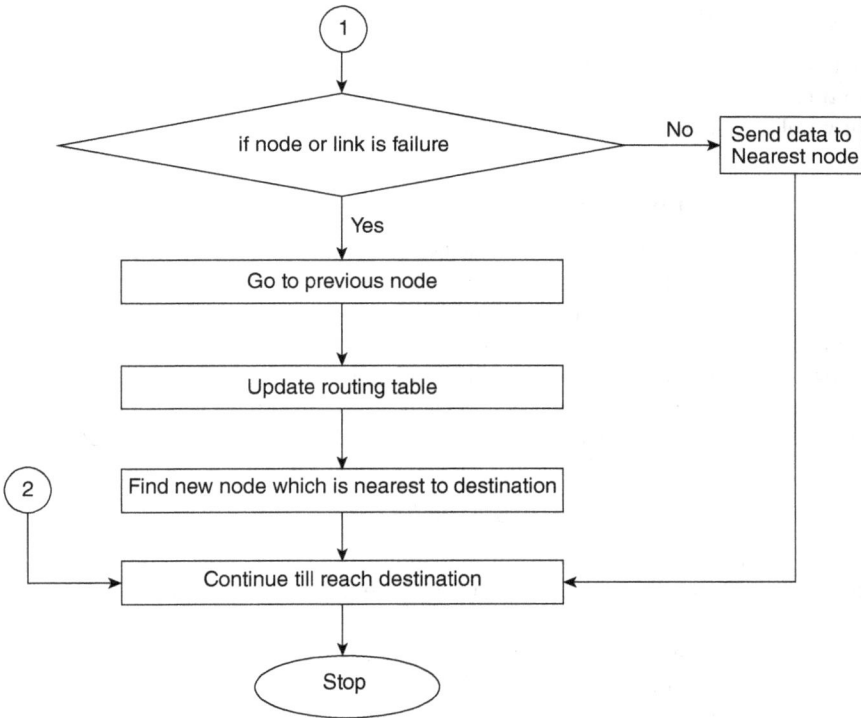

FIGURE 1.7 (Continued)

case, send REQ ANT data via that path. The neighbors' entries in the routing table will be updated. In all cases, repeat this process until the data reaches its destination.

1.3.6 ROUTE ESTIMATION BY FUZZY LOGIC

In this routing protocol, route estimation is also possible by fuzzy logic, which helps to select an optimal as well as feasible route. Fuzzy logic works with several types of membership functions that help protect imprecise information as well as uncertainties. It helps in dealing with linguistic variables based on the nature of membership function. The proposed method is the fusion of three nature-inspired techniques, such as artificial neural network, particle swarm optimization, and ant colony optimization. The combination of the stated methods helps in modeling the network as enhance to network lifetime. But sometimes, it creates uncertainties during fluctuation of several types of information or network parameters. So network parameters map into some linguistic variables by membership function. It helps to create fuzzy variable and model the network in such a way that it easily selects a feasible as well as optimal route.

1.4   EXPERIMENTAL EVALUATION AND ANALYSIS

The suggested scheme's performance is evaluated in several simulation scenarios in this section. The simulations were run in OMNET++, which supports a variety of wireless network routing technologies.

1.4.1   SIMULATION ENVIRONMENT

The network simulator OMNET++ is a discrete, event-driven network simulator that is object-oriented. It is an extremely valuable tool for network simulations. Because of its simplicity and adaptability, the OMNET++ network simulator has garnered tremendous appeal among research community participants. The network simulation allows simulation scripts, also known as simulation scenarios, to be produced quickly in a language, while still relying on more advanced capabilities with C++ code. The proposed protocol is compared with existing protocols such as PSO and ACO.

1.4.2   CONSTRUCT A WIRELESS SENSOR NETWORK SYSTEM

Many sensor nodes are deploying in regions and connect all nodes with each other via a wireless network. Construct a wireless sensor network system and define the routing path. This is shown in Figure 1.8.

Figure 1.9 shows the routing path following source node 0 to node destination 15. Firstly, define the source and destination node; after this source node, find out the neighboring node information from the routing table. Compare all nodes with their respective parameters and find out the nearest to destination. The data packet is forwarded to the corresponding node which is the shortest distance from the destination node. Repeat this process until the data packet reaches the destination. Figure 1.10 shows throughput rate with respect to PSO and ACO protocol, and Figure 1.11 shows delay with respect to PSO and ACO protocols.

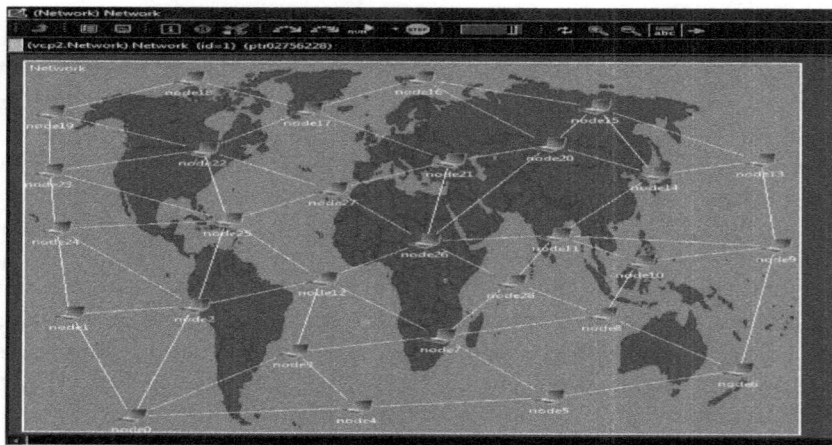

FIGURE 1.8   Deployment of sensor networks.

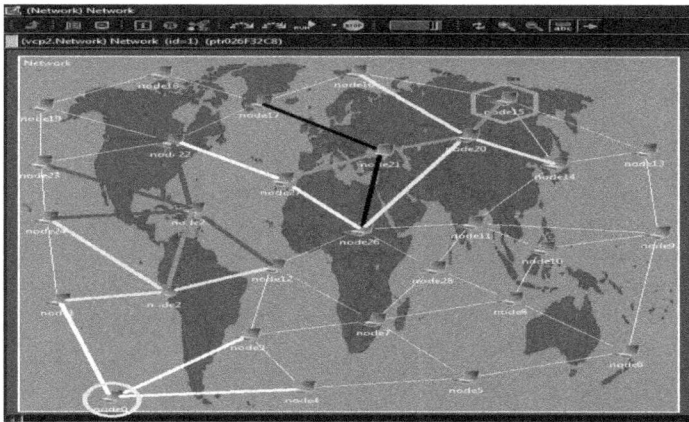

FIGURE 1.9 Routing path from source node 0 to node destination 15.

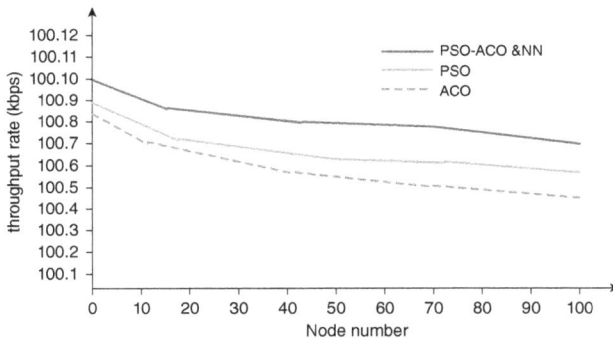

FIGURE 1.10 Throughput rate with respect to PSO and ACO and PSO-ACO with NN protocol.

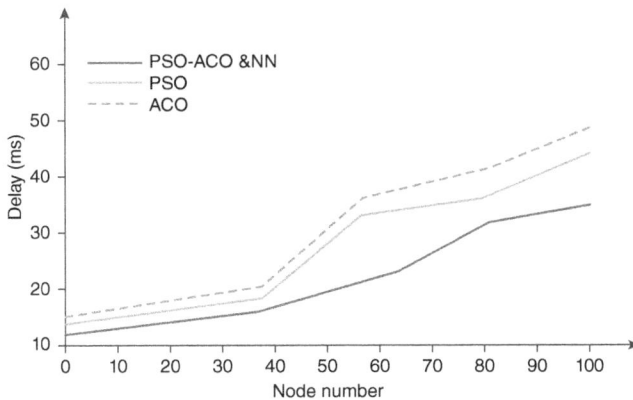

FIGURE 1.11 Delay with respect to PSO and ACO and PSO-ACO with NN protocols.

1.5 CONCLUSION

For WSN, a routing strategy called PSO-ACO and NN has been proposed in this article. The suggested scheme's performance was compared to that of two other reactive protocols, PSO and ACO. According to the principle, PSO-ACO and NN deliver superior results as compared to PSO and ACO. During a route error, the decision maker works based on its decision capacity and decides an alternate way based on route-selection activities. The implementation and analysis of the proposed method shows it outperforms in terms of several metrics, such as throughput, delay, and so on, as evidenced by the simulation results. This paper is useful in the field of wireless sensor networks for optimizing routing paths.

1.6 REFERENCES

[1] Temene, N., Sergiou, C., Georgiou, C., & Vassiliou, V. (2022). A survey on mobility in wireless sensor networks. *Ad Hoc Networks*, 125, 102726, https://doi.org/10.1016/j.adhoc.2021.102726.

[2] Shankar, V. (2021). Contemporary secured target locality in wireless sensor networks. *Global Transitions Proceedings*, 2(2), 194–198.

[3] Binh, H. T. T., Hanh, N. T., Van Quan, L., & Dey, N. (2018). Improved cuckoo search and chaotic flower pollination optimization algorithm for maximizing area coverage in wireless sensor networks. *Neural Computing and Applications*, 30(7), 2305–2317.

[4] Honarparvar, S., Malek, M., Saeedi, S., & Liang, S. (2021). IPAWL: An integrated power aware Wireless sensor network and Location-Based social network for incidence reporting. International *Journal of Applied Earth Observation and Geoinformation*, 104, 102540.

[5] Binh, H. T. T., Hanh, N. T., Nghia, N. D., & Dey, N. (2020). Metaheuristics for maximization of obstacles constrained area coverage in heterogeneous wireless sensor networks. *Applied Soft Computing*, 86, 105939.

[6] Pahar, M., Klopper, M., Warren, R., & Niesler, T. (2021). COVID-19 cough classification using machine learning and global smartphone recordings. *Computers in Biology and Medicine*, 104572, https://doi.org/10.1016/j.compbiomed.2021.104572.

[7] Parvin, J. R., & Vasanthanayaki, C. (2019). Particle swarm optimization-based energy efficient target tracking in wireless sensor network. *Measurement*, 147, 106882.

[8] Singh, A. K., Pamula, R., & Srivastava, G. (2022). An adaptive energy aware DTN-based communication layer for cyber-physical systems. *Sustainable Computing: Informatics and Systems*, 100657, https://doi.org/10.1016/j.suscom.2022.100657.

[9] De, D., Mukherjee, A., Das, S. K., & Dey, N. (2020). Wireless sensor network: Applications, challenges, and algorithms. In *Nature Inspired Computing for Wireless Sensor Networks* (pp. 1–18). Springer, Singapore.

[10] Zhu, J., Jiang, D., Ba, S., & Zhang, Y. (2017). A game-theoretic power control mechanism based on hidden Markov model in cognitive wireless sensor network with imperfect information. *Neurocomputing*, 220, 76–83.

[11] Gulati, K., Kumar, S. S., Boddu, R. S. K., Sarvakar, K., Sharma, D. K., & Nomani, M. Z. M. (2021). Comparative analysis of machine learning-based classification models using sentiment classification of tweets related to COVID-19 pandemic. *Materials Today: Proceedings*, https://doi.org/10.1016/j.matpr.2021.04.364.

[12] Das, S. K., Samanta, S., Dey, N., Patel, B. S., & Hassanien, A. E. (Eds.). (2021). *Architectural Wireless Networks Solutions and Security Issues*. Springer, Singapore.

[13] Singh, A. K., Pamula, R., Jain, P. K., & Srivastava, G. (2021). An efficient vehicular-relay selection scheme for vehicular communication. *Soft Computing*, 1–17.

[14] Esposito, C., & Choi, C. (2017). Signaling game based strategy for secure positioning in wireless sensor networks. *Pervasive and Mobile Computing*, 40, 611–627.

[15] de Fátima Cobre, A., Stremel, D. P., Noleto, G. R., Fachi, M. M., Surek, M., Wiens, A., & Pontarolo, R. (2021). Diagnosis and prediction of COVID-19 severity: Can biochemical tests and machine learning be used as prognostic indicators? *Computers in Biology and Medicine*, 104531.

[16] Das, S. K., Das, S. P., Dey, N., & Hassanien, A. E. (Eds.). (2021). *Machine Learning Algorithms for Industrial Applications*. Springer, Switzerland.

[17] Singh, A. K., & Pamula, R. (2021). An efficient and intelligent routing strategy for vehicular delay tolerant networks. *Wireless Networks*, 27(1), 383–400.

[18] Phoemphon, S., So-In, C., & Niyato, D. T. (2018). A hybrid model using fuzzy logic and an extreme learning machine with vector particle swarm optimization for wireless sensor network localization. *Applied Soft Computing*, 65, 101–120.

[19] Kassania, S. H., Kassanib, P. H., Wesolowskic, M. J., Schneidera, K. A., & Detersa, R. (2021). Automatic detection of coronavirus disease (COVID-19) in X-ray and CT images: A machine learning based approach. *Biocybernetics and Biomedical Engineering*, 41(3), 867–879.

[20] De, D., Mukherjee, A., Das, S. K., & Dey, N. (Eds.). (2020). *Nature Inspired Computing for Wireless Sensor Networks*. Springer, Singapore.

[21] Singh, A. K., & Pamula, R. Vehicular delay tolerant network based communication using machine learning classifiers. *Architectural Wireless Networks Solutions and Security Issues*, 195, https://doi.org/10.1007/978-981-16-0386-0_11

[22] Nuo, W., Qiang, G. U. O., Ming-Lei, S. H. U., LÜ, J. L., & Ming, Y. A. N. G. (2012). Three-dimensional localization algorithm of wireless sensor networks base on particle swarm optimization. *The Journal of China Universities of Posts and Telecommunications*, 19, 7–12.

[23] Tamal, M., Alshammari, M., Alabdullah, M., Hourani, R., Alola, H. A., & Hegazi, T. M. (2021). An integrated framework with machine learning and radiomics for accurate and rapid early diagnosis of COVID-19 from chest X-ray. *Expert Systems with Applications*, 180, 115152, https://doi.org/10.1016/j.eswa.2021.115152.

[24] Das, S. K., Samanta, S., Dey, N., & Kumar, R. (Eds.). (2020). *Design Frameworks for Wireless Networks*. Springer, Singapore.

[25] Singh, A. K., Bera, T., & Pamula, R. (2018, March). PRCP: Packet replication control based prophet routing strategy for delay tolerant network. In *2018 4th International Conference on Recent Advances in Information Technology (RAIT)* (pp. 1–5). IEEE, Dhanbad, India.

[26] Lu, Y., Chen, J., Comsa, I., Kuonen, P., & Hirsbrunner, B. (2014). Construction of data aggregation tree for multi-objectives in wireless sensor networks through jump particle swarm optimization. *Procedia Computer Science*, 35, 73–82.

[27] Lam, C., Siefkas, A., Zelin, N. S., Barnes, G., Dellinger, R. P., Vincent, J. L., & Das, R. (2021). Machine learning as a precision-medicine approach to prescribing COVID-19 pharmacotherapy with remdesivir or corticosteroids. *Clinical Therapeutics*, https://doi.org/10.1016/j.clinthera.2021.03.016

[28] Das, S. K., Maheswari, V., & Sharma, A. (2021). Wireless networks: Applications, challenges, and security issues. In *Architectural Wireless Networks Solutions and Security Issues* (pp. 1–10). Springer, Singapore.

[29] Tang, M., Xin, Y., Li, J., & Zhai, J. (2013). Nonconvex resource control and lifetime optimization in wireless video sensor networks based on chaotic particle swarm optimization. *Applied Soft Computing*, 13(7), 3273–3284.

[30] Weng, F., Zhang, H., & Yang, C. (2021). Volatility forecasting of crude oil futures based on a genetic algorithm regularization online extreme learning machine with a forgetting factor: The role of news during the COVID-19 pandemic. *Resources Policy*, 73, 102148, https://doi.org/10.1016/j.resourpol.2021.102148.

[31] Das, S. K. (2021). Smart design and its applications: Challenges and techniques. *Nature-Inspired Computing for Smart Application Design*, 1.

[32] Salehian, S., & Subraminiam, S. K. (2015). Unequal clustering by improved particle swarm optimization in wireless sensor network. *Procedia Computer Science*, 62, 403–409.

[33] Huang, C., Wang, M., Rafaqat, W., Shabbir, S., Lian, L., Zhang, J., & Song, W. (2021). Data-driven test strategy for COVID-19 using machine learning: A study in Lahore, Pakistan. *Socio-economic Planning Sciences*, 101091, https://doi.org/10.1016/j.seps.2021.101091.

[34] Das, S. K., Dao, T. P., & Perumal, T. (Eds.). (2021). *Nature-Inspired Computing for Smart Application Design*. Springer Nature, Singapore.

[35] Ahmed, H. I., Nasr, A. A., Abdel-Mageid, S. M., & Aslan, H. K. (2021). DADEM: Distributed attack detection model based on big data analytics for the enhancement of the security of internet of things (IoT). *International Journal of Ambient Computing and Intelligence (IJACI)*, 12(1), 114–139.

[36] Sholla, S., Mir, R. N., & Chishti, M. A. (2021). A fuzzy logic-based method for incorporating ethics in the internet of things. *International Journal of Ambient Computing and Intelligence (IJACI)*, 12(3), 98–122.

[37] Balusa, B. C., & Gorai, A. K. (2021). Development of fuzzy pattern recognition model for underground metal mining method selection. *International Journal of Ambient Computing and Intelligence (IJACI)*, 12(4), 64–78.

[38] Cao, B., Zhao, J., Lv, Z., Liu, X., Kang, X., & Yang, S. (2018). Deployment optimization for 3D industrial wireless sensor networks based on particle swarm optimizers with distributed parallelism. *Journal of Network and Computer Applications*, 103, 225–238.

2 Fuzzy Inference-Based Optimal Route-Selection Technique in Wireless Ad Hoc Network

Santosh Kumar Das, Nikhil Patra, Biswa Ranjan Das, and Aditya Sharma

CONTENTS

2.1 Introduction 23
2.2 Related Works... 24
2.3 Proposed Method.. 28
    2.3.1 Network Initialization.. 28
    2.3.2 Parameter Formulation 30
    2.3.3 Fuzzification of Parameters 30
    2.3.4 Evaluation of Optimal Route.. 32
2.4 Simulation and Analysis.. 35
2.5 Conclusions.. 38
2.6 References... 39

2.1 INTRODUCTION

There are several variations of wireless network. Among all variations, one of the most popular variations is wireless ad hoc network, or WANET. It is one of the popular wireless networks which is designed for temporary purpose. In this network, there is no base station or router that helps in communicating one node to another node. Some of the variations are named as mobile ad hoc network, or MANET; vehicular ad hoc network, or VANET; etc. [1, 2]. It is used in several operations and applications in terms of management. Some of the features are mentioned in this section as:

    a. Dynamic topology
    b. Randomness

DOI: 10.1201/b23138-3

    c. Infrastructureless
    d. Parameter variation
    e. Scalability

There are several benefits and applications of an ad hoc network as well as a wireless sensor network. Each of the application is based on a quality-of-service management system based on customer applications. Some of the variations of the network are mentioned in this section as ad hoc on-demand multipath distance vector, or AOMDV; enhanced ant dynamic source routing, or EADSR; ad hoc on-demand distance vector routing, or AODV; etc. [3, 4]. The combination of all routing systems and models helps to communicate several operation and routing systems. There are some parameters and packet information of the network, such as route request or route reply or route error. The combination of these three packet information helps to model several data packet systems and their management based on variation. It causes several limitations and constraints, such as limited bandwidth, limited energy constraints, overhead, etc. So the current paper is based on an inference management system based on optimal route selection for wireless ad hoc network. The route is selected based on the parameter of optimality for efficient route selection.

The remaining parts of the article are divided, as the next part describes the details of existing works. The next part describes a detailed information about the proposed method. The next part analyzes the simulation analysis part, and the last section describes the conclusion of the paper.

2.2   RELATED WORKS

In this section, several works are discussed with the context of existing works based on wireless ad hoc network as well as wireless sensor network. The combination of both types of literature helps to model the application to be efficient with the context of real-life issues. Keerthika and Shanmugapriya [5] designed a method that combines passive and active attacks for the purpose of illustration. This illustration is based on some countermeasure system that helps in vulnerabilities system. It helps in modeling the application based on environment analysis that helps to deal with a protection system based on commercial analysis. The communication of the system is based on the deployment of some challenges along with issues. It helps to make a defensive analysis based on vulnerability analysis based on some factors of information. Wan and Chen [6] designed a strategy for energy analysis and mechanism for harvesting analysis. The work is based on the WSN purpose of modeling. It helps in modeling several cooperative analyses for node analysis. It defines some probability based on relay node detection. The main purpose of this analysis to solve network performance based on certain factors. It helps in modeling the application and saves the actual energy. It uses mathematical modeling for analyzing data and parameters. It helps in enhancing the energy based on solar energy system and its cooperation. N. Dey [7] proposed a method for a nature-inspired optimization system that helps in modeling several variant information. The work is based on several applications and modeling that helps in modeling several information systems. There are several works and applications combined in this application system that help to model nonlinear optimization efficiently and effectively. The contribution of this book helps in modeling several optimizations with the context of real-life application. It helps to model several sources of information based on multiapplication systems based on power management. S. K. Das [8] proposed a method for the purpose of application design system and its management. It helps in several application management systems that help in dealing with several challenges and issues management. It helps to model several information systems based on smart application systems. It helps to model several security management systems based on emergency management and application systems. It helps in giving new guidelines that help model several issues in terms of solution. This solution helps in modeling several services based on real-life application management. It helps in several monitoring and application management based on emergency and security modeling system based on services. Singh et al. [9] proposed an adaptive method for the purpose of energy awareness system that is used for communication purposes. This method is used for physical communication systems that are based on a delay-tolerant system. It helps in modeling applications based on a cyber-physical system that is used for wireless sensor network. It helps to model applications based on delay-tolerant and prediction systems. The application is based on mobile applications for operating several operations. The work is simulated based on network a environment for predicting network lifetime. Misra et al. [10] designed an implementation method based on the fusion of FPGA and NLOS. The work is based on distance analysis and its estimation system. It helps in modeling several applications that help elderly modeling. The work is designed for the purpose of an indoor system that helps in WSN. It helps in location analysis based on the ZigBee network. The work uses programmable gate array system and its modeling. This model uses artificial neural networks to estimate different errors and improve network lifetime. It uses a hybridization method for modeling several complexities based on suitable analysis. Wang and Hu [11] designed a hole-detection method for handling several issues based on WSN. The network is based on a clustering method and an algorithm that uses some gap coverage analysis. It helps to analyze multihop management systems for rational deployment. It helps distance parameter systems and vulnerability detection, which help in coverage and its parameter modeling. It overcomes the limitation of several determination systems for edge node modeling. It helps in determining random walk connection and its management. Singh et al. [12] designed an efficient modeling method for the purpose of helping in vehicular communication. The work is based on communication purposes and modeling. It helps several applications that help in modeling several vehicular relay managements for an efficient network communication. The work is based on utilization purposes that help model several potential system managements. It helps in modeling several relay applications and network modeling. It helps in predicting several vehicular communications based on selfish note prediction. It helps in strategy management for tracking network performance. Das et al. [13] designed a system and modeling for the purpose of wireless network and application. It helps to model several information based on service management and application modeling. The work is based on several information and management systems, such as energy resource management and modeling. It helps deal with several security and privacy management systems that help in design and its enhancement. The work is based on troubleshooting and an automation system for network lifetime management and its enhancement. It gives several protocols for the purpose of design perspective and modeling. Temene et al. [14] illustrated a survey based on mobility analysis and prediction for WSN. The work is based on IoT and WSN both for detailed illustration. It helps in modeling several mobile nodes. There are several mobile nodes that play different roles, such as the sink node, mobile node, source node, etc. The combination of all nodes helps in modeling several congestions and the related mitigations. It helps in the predecessor analysis of IoT, which helps in several directions. The work helps in modeling several evaluations based on different algorithms. Yousefpoor et al. [15] designed a secure method for WSN as a review paper that helps model several issues in the network. The work is based on a data aggregation method that helps in the reduction of the attack in the system. It helps in countermeasures for several issues with the context of attack measurement. This review is also based on industrial internet of things system and its modeling. It helps in the management of several issues with the context of solution measurement. It helps save energy and increase security of the system based on an authentication system. Singh and Pamula [16] designed a method for the purpose of intelligent communication and route modeling. The work is based on a vehicular communication system based on strategy management. It helps based on a strategy behavior modeling and analysis that helps enhance network lifetime. The application is deployed in several purposes based on protocol management. It helps track analysis in novel behavior based on delay-tolerant management. It helps to utilize and analyze the system based on vehicular communication. It helps outperform the results based on several variations. De et al. [17] designed a book for the purpose of wireless sensor network. It helps model several applications based on services and management. It helps deal with several information systems and in the management of key area management. The work is based on nature-inspired application and computing that helps model several issues. It helps implement several application and computation information systems. Information of this book is distributed in the form of bio- and nature-inspired systems. It helps model and design several applications for the purpose of single-objective and multiobjective optimization systems. Zhang and Mao [18] designed a multifactor system for authentic purposes. The work is based on a protocol system that helps model the application. It helps model several recognitions to exercise the physical system. It is based on the ZigBee network system, which helps model several scope identifications. It helps in security analysis and recognition of several applications based on component analysis and its management. It helps model several information based on radio frequency analysis. It helps in designing the system based on security analysis for connection management of the network. Huanan et al. [19] designed a security-based application system for the purpose of wireless sensor network. It helps model several systems for handling several intrusion and detection systems. It helps model several systems based on a foundation of network. It helps in the study of the system for reducing several threats. It helps in modeling some analysis and emphasizes communication and modeling. It helps in handling several security systems for designing some analysis and its modeling. Singh and Pamula [20] designed a delay management for the purpose of a communications-based system. The work is based on a machine learning–based application that helps in several purposes. The work is based on several classifier model of analysis based on solutions for vehicular network. It helps in dealing with several network metrics based on certain parameters of the network. The work is based on strategy management for handling the inadequate design of the system. It helps outperform the result based on the consideration of system modeling. Das et al. [21] designed a book for the purpose of an architectural solution system based on wireless network. This service is not based only on network; it is also based on several systems and information based on the architecture of the network. It helps in modifications based on the architecture of the system. It helps model several issues of the network. Several issues are used and deal with the system. Some of the issues are mentioned, such as energy efficiency system, network lifetime system, resource management, data aggregation system, etc. It helps model several solutions and in the security management of wireless networks. Prasad and Shivashankar [22] designed an enhanced protocol system for ad hoc networks. In this system, the network is based on mobile ad hoc network that helps design routing systems. It helps in the management of several challenges and routing information systems as a way of management and its analysis. It helps manage several source nodes and their information systems based on target node analysis. The work is based on policy management that helps communicate the system efficiently. It helps manage systems based on an autonomous system that helps manage several network lifetimes. Singh and Sharma [23] designed a process for a routing system that helps model mobile ad hoc networks. It is based on an optimization process that helps in the resilience of the operation. It also helps monitor the environment based on vehicular ad hoc networks. The work helps model several device-to-device communication based on the requirement. The outcome of this application is deployed in several areas based on service management. The work is optimized based on nature-inspired optimization for handling several eliminations. Singh et al. [24] designed a method for the purpose of strategy management that is based on delay analysis. It helps maintain several network managements based on some replication analyses. It helps in dealing with and managing several packet control management systems. It helps manage several data delivery and resource management for the purpose of communication systems handling. The work is based on a routing analysis that helps deal with and manage intermediate node analysis systems. The work deals with several assists for the purpose of replication management. This replication is based on a control system and a parameter dealing system. Das et al. [25] designed an application management illustration for the purpose of wireless network systems and services. It helps model several applications based on different security and challenge system management. It helps in issues management based on several parts of the communication system. It helps adopt several solutions based on some higher analysis and management. The work is based on complexity management, which helps model several solutions with the context of management based on different variations of the wireless network. Anand et al. [26] designed a framework system for managing several applications based on multicast service. It helps manage several protocol systems based on service management. The work is based on dependability analysis and productivity management that helps in single transmission. The work helps model several retreating systems to manage transparency. It helps deal with several confident and legitimate analyses to manage node along with network information. It helps manage several intrusions and increases network lifetime. Tahir et al. [27] designed a clustering system to manage several communication systems based on peer-to-peer network and its management. The work helps deal with several overlay management based on application systems. The clustering system of the network is based on multiple dimensional analyses. It helps manage several linkages and its analysis, which helps in lookup management. It helps decrease several complexities, such as overhead, computation system, error, etc. [28–30]. Finally, it helps in modeling several environment analyses based on path management.

2.3 PROPOSED METHOD

In this section, the proposed method is described briefly. The proposed method is divided into four phases illustrated as follows. Some of the abbreviations and their details are shown in Table 2.1.

2.3.1 NETWORK INITIALIZATION

This section is based on network formulation by gram G that contains two basic elements, such as V and E. In this formulation, V is the set of ad hoc nodes and

TABLE 2.1
List of Abbreviation with Description

Abbreviation	Full form
CP	Control Packet
E	Energy
H	High
HC	Hop Count
L	Low
LV	Linguistic Variable
M	Medium
MFs	Membership Functions
RE	Residual Energy

TABLE 2.2

Environment Information

Node ID	RE	CP	X-Coordinate	Y-Coordinate	Radio Range
n_1	E1	C_1	X1	Y1	R1
...	...	...	...	...	...
...	...	...	...	...	...
n_t	E_t	C_t	X_t	Y_t	R_t

E indicates a set of paths between two nodes. It helps in transmitting data from one node to another node. The set of nodes are shown in equation 1. In this equation, t indicates the number of nodes, where it varies based on rounds. The set of nodes t consists of three parameters, shown in equation 2, where E is the residual energy, C is the control packet, and H is the hop count. Description of the crucial parameter is shown in equation 3.

$$N = \{n_1, n_2, n_3, \ldots \ldots \ldots \ldots, n_t\} \tag{1}$$

$$P = \{E, CP, HC\} \tag{2}$$

$$E = E_{initial} - E_{used} \tag{3}$$

In equation 3, $E_{initial}$ is the initial energy, which is assigned randomly during node deployment. So before the start of the transmission, E is the equivalent to $E_{initial}$ and E_{used}, shown in equation 4. Here, E_{used} is the combination of energy used during sending and receiving the data packet.

$$E_{used} = E_{send} + E_{received} \tag{4}$$

Information on the environment of the network created in the XY plane is given in Table 2.2.

In Table 2.2, t is the number of nodes, which varies in each runtime environment. This value also maintains the other column values, such as residual energy, control packet, X and Y coordinate, and radio range. The value of residual energy lies between 0 and 500 J. The range of control packet is in between 0 and 100. The XY coordinate range is between 0 and 300. The radio range is considered as 200 m.

2.3.2 PARAMETER FORMULATION

In this subsection, the parameters of the network are formulated. In this phase, with the help of coordinates, an adjacency matrix is created that shows the initial Euclidean distance of each node to all other nodes shown in Table 2.3. The distance which generated is 200 (i.e. radio range), represented with a dummy

TABLE 2.3
Adjacency Matrix for Nodes

D_{11}	$\cdots$	$\cdots$	D_{1t}
$\cdots$	$\cdots$	$\cdots$	$\cdots$
$\cdots$	$\cdots$	$\cdots$	$\cdots$
$\cdots$	$\cdots$	$\cdots$	$\cdots$
D_{t1}	$\cdots$	$\cdots$	D_{tt}

TABLE 2.4
Possible Paths between Source and Destination Nodes

$Path_1 = [n_1, n_2, \ldots .n_t]$	$H_1 = t\text{-}2$	$E_1 = e1$	$C_1 = c1$
$\cdots$	$\cdots$	$\cdots$	$\cdots$
$\cdots$	$\cdots$	$\cdots$	$\cdots$
$Path_{max} = [n_1, n_2, \ldots .n_t]$	$H_{max} = t\text{-}2$	$E_{max} = e_{max}$	$C_{max} = c_{max}$

value (i.e. 99999), and the self-loop distance is constant (i.e. 0). So all the nonzero values are the actual distance between nodes n_i and n_{i+1}, which satisfies the radio signal.

In Table 2.3, the value of D_{ij} may be 0 or 99999 or any other within radio range based on the randomized situation. Source and destination nodes are any nodes within t nodes. Then, between source and destination nodes, several possible paths are analyzed, as shown in Table 2.4.

In Table 2.4, all possible paths are calculated between the source and destination nodes, including three parameter evaluations. The number "max" shows total number of paths between the source and destination nodes. It varies in each runtime environment based on randomization. The values of each path show node numbers, including source and destination nodes, but here the maximum possible node is t. So here t is given at last. For each path, total hop count is calculated in the second column, which can be a maximum of "t-2" because two nodes are excluded due to the source and destination nodes. In the third column, the total residual energy consumed is calculated for each path, which varies based on the considered residual energy. In the fourth column, the total control packet is used for each path calculated, which varies based on the considered control packet.

2.3.3 FUZZIFICATION OF PARAMETERS

In this phase, fuzzification is used for the parameters to control uncertainty and reduce impreciseness. In fuzzy logic, there are five phases or steps for controlling the impreciseness, given as Crisp Data → Fuzzification → Inference Engine with Rule Base → Defuzzification → Crisp Data. Initially, crisp data is used for operation, then it fuzzifies for converting crisp data to fuzzy data. The converted fuzzy data is used for generating a rule-based system. After analyzing the rule-based system, it produces a fuzzy value. Finally, these fuzzy value again is converted into crisp data with the help of a defuzzification method. The crisp value of energy parameter is 500, the control packet is 100, and the hop count is a dynamic value which is generated during runtime. In this fuzzification system, the triangular membership function is used for each parameter. The membership functions of input parameters are shown in Tables 2.5 to 2.7, and their fuzzy set equations are shown in equations 5 to 7. These fuzzy sets contain ten elements for operation of route evaluation.

$$E = RE(e, u_e(i)) \tag{5}$$

$$C = CP(c, u_c(i)) \tag{6}$$

$$H = HC(h, u_h(i)) \tag{7}$$

TABLE 2.5
Membership Function of Energy

LV	Notation	Range
L	E_L	$[E_{L-}, E_{L+}]$
M	E_M	$[E_{M-}, E_{M+}]$
H	E_H	$[E_{H-}, E_{H+}]$

TABLE 2.6
Membership Function of Control Packet

LV	Notation	Range
L	C_L	$[C_{L-}, C_{L+}]$
M	C_M	$[C_{M-}, C_{M+}]$
H	C_H	$[C_{H-}, C_{H+}]$

TABLE 2.7
Membership Function of Hop Count

LV	Notation	Range
L	H_L	$[H_{L-}, H_{L+}]$
M	H_M	$[H_{M-}, H_{M+}]$
H	H_H	$[H_{H-}, H_{H+}]$

2.3.4 Evaluation of Optimal Route

In this subsection, final route is decided based on the aforementioned input parameters. In this work, the output parameters are named as stability (i.e., S). It is the parameter that will decide the best path out of all the paths from source to destination. Among the three input parameters, energy is the very important or crucial parameter. So energy is mapped with two other input parameters, such as control packet and hop count, separately. Hence, two rule-based systems (RBS) are created, as shown in equations 8 and 9. The membership functions of the two RBS are shown in Tables 2.8 and 2.9.

$$RBS1 = (E \times C) \tag{8}$$

$$RBS2 = (E \times H) \tag{9}$$

TABLE 2.8
Membership Function for RBS1

LV	Notation	Stability1
Poor	P	RBS_{P1}
Very Bad	VB	RBS_{VB1}
Bad	B	RBS_{B1}
Almost Bad	AB	RBS_{AB1}
Medium	M	RBS_{M1}
Quite Good	QG	RBS_{QG1}
Good	G	RBS_{G1}
Very Good	VG	RBS_{VG1}
Excellent	E	RBS_{E1}

TABLE 2.9
Membership Function for RBS2

LV	Notation	Stability2
Poor	P	RBS_{P2}
Very Bad	VB	RBS_{VB2}
Bad	B	RBS_{B2}
Almost Bad	AB	RBS_{AB2}
Medium	M	RBS_{M2}
Quite Good	QG	RBS_{QG2}
Good	G	RBS_{G2}
Very Good	VG	RBS_{VG2}
Excellent	E	RBS_{E2}

In Tables 2.8 and 2.9, the total variable is 9, because in each stability set is 3 × 3, i.e., first for residual energy and control packet, and second for residual energy and hop count. The linguistic variables for both stabilities are arranged based on a chronological order. Tables 2.10 and 2.11 shown a rule-based system for both stabilities. Tables 2.12 and 2.13 show for both fuzzy rule-based matrix. In the first table, the row is labeled as residual energy and the column is labeled as control packet. The different cells are labeled by each linguistic variable for stability 1. In the second table, the row is labeled as residual energy, and the column is labeled as hop count, while the different cells are labeled by each linguistic variable for stability 2.

The parameter residual energy is directly proportional to stability because if energy is high, then network lifetime is also high, so stability of the network is also better. But two parameters, hop count and control packet, are inversely

TABLE 2.10
Rule-Based System for Stability 1

Rule No.	Rule Name
1	If E is L and CP is L, then stability1 is good.
2	If E is L and CP is M, then stability1 is almost good.
3	If E is L and CP is H, then stability1 is poor.
4	If E is M and CP is L, then stability1 is very good.
5	If E is M and CP is M, then stability1 is medium.
6	If E is M and CP is H, then stability1 is very bad.
7	If E is H and CP is L, then stability1 is excellent.
8	If E is H and CP is M, then stability1 is quite good.
9	If E is H and CP is H, then stability1 is bad.

TABLE 2.11
Rule-Based System for Stability 2

Rule No.	Rule Name
1	If E is L and HC is L, then Stability2 is good.
2	If E is L and HC is M, then Stability2 is almost good.
3	If E is L and HC is H, then Stability2 is poor.
4	If E is M and HC is L, then Stability2 is very good.
5	If E is M and HC is M, then Stability2 is medium.
6	If E is M and HC is H, then Stability2 is very bad.
7	If E is H and HC is L, then Stability2 is excellent.
8	If E is H and HC is M, then Stability2 is quite good.
9	If E is H and HC is H, then Stability2 is bad.

TABLE 2.12
Fuzzy Rule-Based Matrix for Stability 1

		Control Packet		
		C_L	C_M	C_H
Energy	E_L	RBS_{G1}	RBS_{AB1}	RBS_{P1}
	E_M	RBS_{VG1}	RBS_{M1}	RBS_{VB1}
	E_H	RBS_{E1}	RBS_{QG1}	RBS_{B1}

TABLE 2.13
Fuzzy Rule-Based Matrix for Stability 2

		Hop Count		
		C_L	C_M	C_H
Energy	E_L	RBS_{G2}	RBS_{AB2}	RBS_{P2}
	E_M	RBS_{VG2}	RBS_{M2}	RBS_{VB2}
	E_H	RBS_{E2}	RBS_{QG2}	RBS_{B2}

TABLE 2.14
Fuzzy Rule-Based Matrix for Stability

		Control Packet vs. Hop Count		
		CH_L	CH_M	CH_H
Energy	E_L	RBS_G	RBS_{AB}	RBS_P
	E_M	RBS_{VG}	RBS_M	RBS_{VB}
	E_H	RBS_E	RBS_{QG}	RBS_B

proportional to stability. Hence, there is need of complement operation. Therefore, the third fuzzy rule matrix is created with the help of both stated fuzzy rule matrices, as shown in Table 2.14.

Equations 5 to 7 contain a total of ten elements among them. Some elements are low, some elements are medium, and some elements are high, so combine all ten elements into three categories, as equations 10 to 12.

$$E_L = Mean(E)_{i=1}^{p} \tag{10}$$

$$E_M = Mean(E)_{j=1}^{q} \tag{11}$$

$$E_H = Mean(E)_{k=1}^{r} \tag{12}$$

Where i, j, and k vary between 1 and p, 1 and q, and 1 and r, and p, q, and r may or may not be equal.

Similarly, the fuzzy set of control packet and hop count is categorized by merging. So here the optimal route is the route number 7, which has high energy but low control packet and hop count, and the next optimal route is route number 4, which has medium energy and low control packet with hop count.

2.4 SIMULATION AND ANALYSIS

The proposed method is simulated into Java programming. Its simulation parameters of the environment are shown in Table 2.15. Figure 2.1 shows the first scenario of the simulation, where the total number of nodes are entered. Figure 2.2 shows network environment with some properties, such as node ID mark of 0 as n_1, 1 as n_2, and so on; residual energy, based on 0 to 500 J; control packet, based on 0 to 100; both coordinate range within 300; and radio range within 200.

Then, with the help of coordinates, prepare an adjacency matrix, as shown in Figure 2.3. It shows the initial Euclidean distance of each node to all other nodes.

TABLE 2.15
Simulation Parameters

Parameter	Description
Windows	Windows 10 Pro
Java programming	12.0.2
MS Office	2013
XY-area	X(0–300) and Y(0–300)
Energy	500
Control packet	1–100
Radio range	200

```
Enter the number of nodes:
5
```

FIGURE 2.1 Total number of node entries.

```
The Details of each nodes:
Node Id Residual Energy  Control Packet   X-Coordinate   Y-Coordinate              Radio Range
0              346               91              293            130                      200
1              312               78              291            221                      200
2              214               84              181             80                      200
3              441               27              206            205                      200
4              445               16              280            234                      200
```

FIGURE 2.2 Network environment.

The distance which is generated more than 200 is represented with a dummy value (i.e. 99999), and the self-loop distance is constant (i.e. 0). So all the nonzero values are the actual distance between two particular nodes that satisfy the radio signal.

Since all the information required about the node is present, here the decision maker first asks for the node ID which acts as source node for the whole runtime operation, and also asks for the node ID which is treated as the destination node in our program. This situation is shown in Figure 2.4.

Taking source and destination as two parameters in Figure 2.4, list out all the possible paths in the given network, as shown in Figure 2.5. Here, along with path, also calculated are the number of hop count for individual path, the energy

```
Now we want the weighted adjacency matrix:

The initial adjacency matrix is:
            0          91         122         114         104
           91           0         178          86          17
          122         178           0         127         183
          114          86         127           0          79
          104          17         183          79           0
```

FIGURE 2.3   Adjacency matrix for node distance.

```
Enter the source node:
0
Enter the destination node:
2
```

FIGURE 2.4   Entering source and destination nodes.

```
Following are all different paths from 0 to 2
path 1= [0, 1, 2]              hops=1          Energy=290        Control Packet=84
path 2= [0, 1, 3, 2]           hops=2          Energy=328        Control Packet=70
path 3= [0, 1, 3, 4, 2]        hops=3          Energy=351        Control Packet=59
path 4= [0, 1, 4, 2]           hops=2          Energy=329        Control Packet=67
path 5= [0, 1, 4, 3, 2]        hops=3          Energy=351        Control Packet=59
path 6= [0, 2]          hops=0          Energy=280        Control Packet=87
path 7= [0, 3, 1, 2]           hops=2          Energy=328        Control Packet=70
path 8= [0, 3, 1, 4, 2]        hops=3          Energy=351        Control Packet=59
path 9= [0, 3, 2]       hops=1          Energy=333        Control Packet=67
path 10= [0, 3, 4, 1, 2]       hops=3          Energy=351        Control Packet=59
path 11= [0, 3, 4, 2]          hops=2          Energy=361        Control Packet=54
path 12= [0, 4, 1, 2]          hops=2          Energy=329        Control Packet=67
path 13= [0, 4, 1, 3, 2]       hops=3          Energy=351        Control Packet=59
path 14= [0, 4, 2]      hops=1          Energy=335        Control Packet=63
path 15= [0, 4, 3, 1, 2]       hops=3          Energy=351        Control Packet=59
path 16= [0, 4, 3, 2]          hops=2          Energy=361        Control Packet=54
```

FIGURE 2.5   Paths between source and destination nodes along with information.

consumed between source to destination, and the average control packet required to complete the signal from source to destination.

Apart from doing this, create a fuzzy set of residual energy where there is a function that will take ten random numbers between 0 and 500, and the function will return its corresponding membership value, represented as RE(e, u_e(i)). Similarly, a fuzzy set of control packet where there is a function that will take ten random numbers between 0 and 100 and the function will return its corresponding membership value, represented as CP(c, u_c(i)). Similarly, a fuzzy set of hop count where there is a function that will take ten random numbers between 0 and max hop count and the function will return its corresponding membership value, represented as HC(h, u_h(i)). Details are shown in Figure 2.6.

The output parameter "stability," which consists of two internal output parameters, as stability 1 and stability 2. Hence, the fuzzy rule matrix for both stabilities is shown in Figure 2.7.

```
Fuzzy Set of Residual Energy:

RE(e,u_e(i))={(0,0.00),(290,0.53),(310,0.32),(342,0.34),(439,0.64),(410,0.95),(347,0.39),(348,0.40),(141,0.43)
,(286,0.57)}

Fuzzy Set of Control Packet:

CP(cp,u_cp(i))={(81,0.90),(51,0.57),(5,0.06),(22,0.24),(38,0.42),(87,0.97),(89,0.99),(25,0.28),(61,0.68)}

Fuzzy Set of HopCount:

HC(h,u_h(i))={(2,1.00),(3,0.50),(5,1.00),(6,0.50),(9,0.50),(10,0.00)}
```

FIGURE 2.6 Fuzzy sets of three input parameters.

```
Rule Base Matrix for Energy and Hop Count
The matrix contains the linguistic variables of Stability of the route
                            Hop Count
                  Low           Medium        High
                -----------------------------------------
        Low     |  Good     |     Bad    |   VV Bad   |
                -----------------------------------------
Energy  Medium  | Very Good | Quite Good |  Very Bad  |
                -----------------------------------------
        High    | Excellent |  Well Done |  Not Bad   |
                -----------------------------------------


Rule Base Matrix for Energy and Control packet
The matrix contains the linguistic variables of Stability of the route
                  Hop Count
                  Low           Medium        High
                -----------------------------------------
        Low     |  Good     |     Bad    |   VV Bad   |
                -----------------------------------------
Energy  Medium  | Very Good | Quite Good |  Very Bad  |
                -----------------------------------------
        High    | Excellent |  Well Done |  Not Bad   |
                -----------------------------------------
```

FIGURE 2.7 Fuzzy rule matrix for stability 1 and stability 2.

```
Values for Residual Energy
[0, 290, 310, 342, 439, 410, 347, 348, 141, 286]
        Low=0.21666666666666667
        Medium=0.4701754385964912
        high=0.5052631578947367
```

FIGURE 2.8   Mean values of energy linguistic variable.

```
Values for Control Packet
[81, 51, 5, 22, 38, 87, 89, 25, 61]
        Low=0.25
        Medium=0.4861111111111111
        high=0.8833333333333333
```

FIGURE 2.9   Mean values of control packet linguistic variable.

```
Values for Hop Count
[2, 3, 5, 6, 9, 10]
        Low=0.75
        Medium=0.6666666666666666
        high=0.3333333333333333
```

FIGURE 2.10   Mean values of hop count linguistic variable.

Meanwhile, the fuzzy set contains ten elements, out of which some are low, some are medium, and some are high in the category of residual energy. Hence, calculate the mean low membership value, mean medium membership value, and mean high membership value. Similarly, do so for control packet and hop count, as shown in Figures 2.8 and 2.10.

2.5 CONCLUSIONS

The proposed method uses fuzzy logic in terms of fuzzy inference systems for handling optimal route by using three input parameters and one output parameter. The first parameter, residual energy, helps boost the network nodes for sending and transmitting the data packets. The second parameter, control packet, helps determine the capacity of the nodes for sending and transmitting the data packet. The third parameter, hop count, is used to cover the distance between the source and destination nodes. The combination of all parameters helps design a fuzzy inference system with a rule-based system. It helps determine the optimal route by reducing uncertainty and ambiguity of the route selection. It precisely evaluates the variations of the input parameters based on the stability output parameter of the network. It not only helps in the selection of the optimal route but also helps in the selection and determination of a feasible route of the network.

2.6 REFERENCES

[1] Prasad, R. (2021). Enhanced energy efficient secure routing protocol for mobile ad-hoc network. *Global Transitions Proceedings*, https://doi.org/10.1016/j.gltp.2021.10.001

[2] Dafalla, M. E. M., Mokhtar, R. A., Saeed, R. A., Alhumyani, H., Abdel-Khalek, S., & Khayyat, M. (2021). An optimized link state routing protocol for real-time application over vehicular ad-hoc network. *Alexandria Engineering Journal*, https://doi.org/10.1016/j.aej.2021.10.013.

[3] Ramya, T. A., Mathana, J. M., Nirmala, R., & Gomathi, R. (2021). Exploration on enhanced quality of services for MANET through modified Lumer and Fai-eta algorithm with modified AODV and DSR protocol. *Materials Today: Proceedings*, https://doi.org/10.1016/j.matpr.2021.05.601.

[4] Saraswat, B. K., Bhardwaj, M., & Pathak, A. (2015). Optimum experimental results of AODV, DSDV & DSR routing protocol in grid environment. *Procedia Computer Science*, 57, 1359–1366.

[5] Keerthika, M., & Shanmugapriya, D. (2021). Wireless sensor networks: Active and passive attacks-vulnerabilities and countermeasures. *Global Transitions Proceedings*, 2(2), 362–367.

[6] Wan, J., & Chen, J. (2022). AHP based relay selection strategy for energy harvesting wireless sensor networks. *Future Generation Computer Systems*, 128, 36–44.

[7] Dey, N. (Ed.). (2021). *Applications of Flower Pollination Algorithm and Its Variants*. Springer, Singapore, ISBN: 978-981-33-6104-1.

[8] Das, S. K. (2021). Smart design and its applications: Challenges and techniques. *Nature-Inspired Computing for Smart Application Design*, 1.

[9] Singh, A. K., Pamula, R., & Srivastava, G. (2022). An adaptive energy aware DTN-based communication layer for cyber-physical systems. *Sustainable Computing: Informatics and Systems*, 100657, https://doi.org/10.1016/j.suscom.2022.100657.

[10] Misra, Y., Krishnaveni, K., & Rajasekaran, A. S. (2022). Implementation of NLOS based FPGA for distance estimation of elderly using indoor wireless sensor networks. *Materials Today: Proceedings*, https://doi.org/10.1016/j.matpr.2022.01.087.

[11] Wang, F., & Hu, H. (2021). Coverage hole detection method of wireless sensor network based on clustering algorithm. *Measurement*, 179, 109449, https://doi.org/10.1016/j.measurement.2021.109449.

[12] Singh, A. K., Pamula, R., Jain, P. K., & Srivastava, G. (2021). An efficient vehicular-relay selection scheme for vehicular communication. *Soft Computing*, 1–17.

[13] Das, S. K., Samanta, S., Dey, N., & Kumar, R. (Eds.). (2020). *Design Frameworks for Wireless Networks*. Springer, Singapore.

[14] Temene, N., Sergiou, C., Georgiou, C., & Vassiliou, V. (2022). A survey on mobility in Wireless Sensor Networks. *Ad Hoc Networks*, 125, 102726, https://doi.org/10.1016/j.adhoc.2021.102726.

[15] Yousefpoor, M. S., Yousefpoor, E., Barati, H., Barati, A., Movaghar, A., & Hosseinzadeh, M. (2021). Secure data aggregation methods and countermeasures against various attacks in wireless sensor networks: A comprehensive review. *Journal of Network and Computer Applications*, 103118, https://doi.org/10.1016/j.jnca.2021.103118.

[16] Singh, A. K., & Pamula, R. (2021). An efficient and intelligent routing strategy for vehicular delay tolerant networks. *Wireless Networks*, 27(1), 383–400.

[17] De, D., Mukherjee, A., Das, S. K., & Dey, N. (Eds.). (2020). *Nature Inspired Computing for Wireless Sensor Networks*. Springer, Singapore.

[18] Zhang, J., & Mao, H. (2021). Multi-factor identity authentication protocol and indoor physical exercise identity recognition in wireless sensor network. *Environmental Technology & Innovation*, 101671, https://doi.org/10.1016/j.eti.2021.101671.

[19] Huanan, Z., Suping, X., & Jiannan, W. (2021). Security and application of wireless sensor network. *Procedia Computer Science*, 183, 486–492.

[20] Singh, A. K., & Pamula, R. (2021). Vehicular delay tolerant network based communication using machine learning classifiers. *Architectural Wireless Networks Solutions and Security Issues*, 195, https://doi.org/10.1007/978-981-16-0386-0_11

[21] Das, S. K., Samanta, S., Dey, N., Patel, B. S., & Hassanien, A. E. (Eds.). (2021). *Architectural Wireless Networks Solutions and Security Issues*. Springer, Singapore.

[22] Prasad, R. (2021). Enhanced energy efficient secure routing protocol for mobile ad-hoc network. *Global Transitions Proceedings*, https://doi.org/10.1016/j.gltp.2021.10.001.

[23] Singh, N. C., & Sharma, A. (2020). Resilience of mobile ad hoc networks to security attacks and optimization of routing process. *Materials Today: Proceedings*, https://doi.org/10.1016/j.matpr.2020.09.622.

[24] Singh, A. K., Bera, T., & Pamula, R. (2018, March). PRCP: Packet replication control based prophet routing strategy for delay tolerant network. In *2018 4th International Conference on Recent Advances in Information Technology (RAIT)* (pp. 1–5). IEEE, Dhanbad, India.

[25] Das, S. K., Maheswari, V., & Sharma, A. (2021). Wireless networks: Applications, challenges, and security issues. In *Architectural Wireless Networks Solutions and Security Issues* (pp. 1–10). Springer, Singapore.

[26] Anand, M., Balaji, N., Bharathiraja, N., & Antonidoss, A. (2021). A controlled framework for reliable multicast routing protocol in mobile ad hoc network. *Materials Today: Proceedings*, https://doi.org/10.1016/j.matpr.2020.10.902.

[27] Tahir, A., Shah, N., Abid, S. A., Khan, W. Z., Bashir, A. K., & Zikria, Y. B. (2021). A three-dimensional clustered peer-to-peer overlay protocol for mobile ad hoc networks. *Computers & Electrical Engineering*, 94, 107364, https://doi.org/10.1016/j.compeleceng.2021.107364.

[28] Ahmed, H. I., Nasr, A. A., Abdel-Mageid, S. M., & Aslan, H. K. (2021). DADEM: Distributed attack detection model based on big data analytics for the enhancement of the security of internet of things (IoT). *International Journal of Ambient Computing and Intelligence (IJACI)*, 12(1), 114–139.

[29] Sholla, S., Mir, R. N., & Chishti, M. A. (2021). A fuzzy logic-based method for incorporating ethics in the internet of things. *International Journal of Ambient Computing and Intelligence (IJACI)*, 12(3), 98–122.

[30] Balusa, B. C., & Gorai, A. K. (2021). Development of fuzzy pattern recognition model for underground metal mining method selection. *International Journal of Ambient Computing and Intelligence (IJACI)*, 12(4), 64–78.

3 Fuzzy-Based Mathematical Model for Optimizing Network Lifetime in MANET

*Manoj Kumar Mandal, Arun Prasad Burnwal,
B. K. Mahatha, and Abhishek Kumar*

CONTENTS

3.1 Introduction 41
3.2 Related Works.. 42
3.3 Proposed Method... 45
3.4 Simulation and Analysis 48
3.5 Conclusion 51
3.6 References.. 52

3.1 INTRODUCTION

In the modern era, most networks are intelligent. The mobile ad hoc network, or MANET, is one of these networks. Its application increases rapidly [1–3]. It is a collection of several mobile nodes where the nature of the mobile node is dynamic. In this network, there is no router. Each node works as a router and helps send or receive the data packet. There are several features of MANET, such as infrastructureless, dynamic topology, energy constraint, etc. The stated features help create several issues, such as interference between nodes and path, communication failure between nodes as well as path, packet loss, overhead, control packet dropping, etc. [4, 5]. There are several issues available for the same purpose and management. Some of the issues are known as overhead, constraint, limited energy, limited mobility, transmission overhead, etc.

The current paper is based on a mathematical model and its several enhancements. It is used for optimization purposes to help enhance network lifetime efficiently. It helps deal with several models based on fuzzy linguistic variable. Fuzzy linguistic variable is mapped with network constraints to model the problem. It uses quadratic programming for the purpose of formulation and its modeling. It helps to model the objective function along with constraints into mathematical modeling and helps tune the model. This tuning is based on network lifetime metrics.

The remainder of the article is divided into sections. The next section discusses several works based on existing methodologies on optimization. The next section deals with the purpose of the proposed method to help model the data. The next section is used for performance analysis and its evaluation. The last section is the conclusion of the article based on the outcome.

3.2   RELATED WORKS

There are several works proposed for the purpose of the mobile ad hoc network. Most of these works are based on optimization purpose by using fuzzy logic and nature-inspired techniques, including intelligent techniques. It helps provide new insight into the areas of intelligent and efficient communication. Some of the works mentioned in this section include that of Prasad and Shivashankar [6], who designed an enhanced protocol system for ad hoc network. In this system, the network is based on mobile ad hoc network that helps design a routing system. It helps manage several challenges and routing information systems for management and its analysis. It helps manage several source nodes and its information system based on target node analysis. The work is based on policy management, which helps communicate the system efficiently. It helps manage the system based on an autonomous system that helps manage several network lifetimes. Singh and Sharma [7] designed a process for routing system that helps model mobile ad hoc network. It is based on an optimization process that helps make the operation resilient. It also helps monitor the environment based on vehicular ad hoc network. The work helps model several device-to-device communication systems based on the requirement. The outcome of this application is deployed in several areas based on service management. The work is optimized based on nature-inspired optimization for handling several eliminations. S. K. Das [8] proposed a method for the purpose of application design system and its management. It helps in several application management systems dealing with several challenges and issue management. It helps model several information systems based on a smart application system. It helps model several security management systems based on an emergency management and application system. It helps give new guidelines that help model several issues in terms of solution. This solution helps model several services based on real-life application management. It helps in several monitoring and application management systems based on emergency and security modeling system based on services. N. Dey et al. [9] designed a method for the purpose of several big data analysis and modeling. It helps model several next-generation information systems. It helps model several intelligence applications based on different types of services. The works of this book are based on the fusion of the internet of things, cloud computing, and several intelligence systems. It helps model new techniques and services for modeling several services based on new and smart applications. It helps model several big data analytics that helps model some internet of things information system. Singh et al. [10] designed an efficient modeling method for the purpose of helping vehicular communication.

The work is based on communication purposes and modeling. It helps several application purposes that helps model several vehicular relay managements for the purpose of efficient network communication. The work is based on the utilization purpose, which helps model several potential system managements. It helps model several relay applications and network modeling. It helps predict several vehicular communications based on selfish note prediction. It helps in strategy management for tracking network performance. Das et al. [11] designed an application management illustration for the purpose of wireless network system and services. It helps model several applications based on different security and challenge system management. It helps in issues management based on several parts of a communication system. It helps adopt several solutions based on some higher analysis and management. The work is based on complexity management, which helps model several solutions with the context of management based on different variation of wireless network. Anand et al. [12] designed a framework system for managing several applications based on multicast service. It helps manage several protocol systems based on service management. The work is based on dependability analysis and productivity management, which help in single transmission. The work helps model several retreating systems to manage transparency. It helps in dealing with several confident and legitimate analyses to manage node along with network information. It helps manage several intrusions and increase network lifetime. Tahir et al. [13] designed a clustering system to manage several communications based on peer-to-peer network and its management. The work helps deal with several overlay management systems based on an application system. The clustering system of the network is based on multiple dimensional analyses. It helps manage several linkages and their analysis to help in lookup management. It helps decrease several complexities, such as overhead, computation system, error, etc. Finally, it helps model several environment analyses based on path management. Singh and Pamula [14] designed a method for the purpose of intelligent communication and routing modeling. The work is based on a vehicular communication system based on strategy management. It helps based on strategy behavior modeling and analysis, which helps enhance network lifetime. The application is deployed in several purposes based on protocol management. It helps track analysis in novel behavior based on delay-tolerant management. It helps utilize and analyze the system based on vehicular communication. It helps outperform the results based on several variations. Das et al. [15] designed a system and model for the purpose of wireless network and its application. It helps model several information based on service management and application modeling. The work is based on several information and management systems, such as energy resource management and modeling. It helps deal with several security and privacy management systems that help in its design and enhancement. The work is based on troubleshooting an automation system for network lifetime management and its enhancement. It gives several protocols for the purpose of design perspective and modeling. Sharma and Kim [16] designed a method for multipath management that helps model several information. It helps model several applications based on routing information, which helps model network. This network is based on application management of mobile ad hoc network.

The work used a bioinspired technique that helps manage some applications based on services. It helps model several constraints, such as low memory management, bandwidth, battery life, etc. The combination of all information helps model several applications along with services to adjust the model of the network. Lee et al. [17] designed a technique that helps model several sharing information systems. The work is based on a military application that uses mobile ad hoc network to help manage some application management systems. The proposed work is named as a cooperative phase with steering system based on relay management. It consists of several relay and destination node management systems based on services. The network also attaches with a cognitive network model to increase the probability of routing. It helps select relay based on several source nodes to increase the performance of the model. Singh and Pamula [18] designed a delay management for the purpose of a communications-based system. The work is based on a machine learning–based application that helps in several purposes. The work is based on several classifier model of analysis based on solutions for the vehicular network. It helps deal with several network metrics based on certain parameters of the network. The work is based on strategy management for handling the inadequate design of the system. It helps outperform the result based on the consideration of the system model. Das et al. [19] designed a book for the purpose of an architectural solution system based on wireless network. This service is not only based on networks but also on several systems and information based on the architecture of the network. It helps in modifications based on the architecture of the system. It helps model several issues of the network. Several issues are used and deal with the system. Some of these issues are mentioned, such as energy efficiency system, network lifetime system, resource management, data aggregation system, etc. It helps model several solutions and in the security management of the wireless network. Keerthika and Shanmugapriya [20] designed a method with a combination of passive and active attacks for the purpose of illustration. This illustration is based on some countermeasure system that helps in vulnerabilities system. It helps model the application based on environment analysis, which helps deal with a protection system based on commercial analysis. The communication of the system is based on the deployment of some challenges along with issues. It helps in defensive analysis based on vulnerability analysis of some factors of information. Wan and Chen [21] designed a strategy for energy analysis and mechanism for harvesting analysis. The work is based on the WSN purpose of modeling. It helps model several cooperative analyses for node analysis. It defines some probability based on relay node detection. The main purpose of this analysis is to solve network performance based on certain factors. It helps model the application and save the actual energy. It uses mathematical modeling for analyzing data and its parameters. It helps enhance the energy based on a solar energy system and its cooperation. Singh et al. [22] proposed an adaptive method for the purpose of an energy-aware system that is used for communication purposes. This method is used for physical communication systems based on a delay-tolerant system. It helps model applications based on a cyber-physical system that is used for wireless sensor network. It helps model applications based on delay-tolerant and prediction systems. The application is based on a mobile application for operating several operations. The work is simulated based on network environment for predicting network lifetime. De et al. [23] designed a book for the purpose of wireless sensor network. It helps model several applications based on services and management. It helps deal with several information systems and in the management of key areas. The work is based on nature-inspired applications and computing that help model several issues. It helps implement several applications and computation information systems. Information of this book is distributed in the form of bio- and nature-inspired systems. It helps model and design several applications for the purpose of single-objective and multiobjective optimization systems. Shaban et al. [24] designed a patient analysis system that uses an inference system and design. It is based on several information that helps model several neural network systems and modeling. The work is based on a fuzzy inference modeling system that helps model some issues of COVID-19. The issue is based on deep learning and fuzzy system modeling. It helps model several network information based on strategy management. It helps validate several cross-analysis and validation that helps in accuracy modeling. It helps model several detections for coronavirus analysis in helping and modeling based on a prevention system. Zahra et al. [25] designed a method for a data-driven system that is based on intelligence system maintenance and information system. The work is based on a URL system information and modeling that helps analyze and design several malicious and phishing information systems. The work is completely based on fuzzy logic modeling. It helps deal with several uncertainty management for handling some pandemic analysis based on learning and security system [26–28]. It helps model several control and unprecedented information systems. The work helps in modeling several cybercriminal information analyses to deal with several ransomware information system for impact analysis.

3.3   PROPOSED METHOD

This section is used for the purpose of presenting the main method analysis and design. The proposed method is used for network lifetime enhancement by using two parameters, such as residual energy and packet delivery ratio. Both parameters are considered as 300 unit and 1,000 unit. Residual energy is also known as energy of the node. The combination of both parameters helps in modeling the network lifetime by using quadratic programming and fuzzy logic. In this section, Tables 3.1 and 3.2 are used to define the fuzzy membership functions of residual energy and the packet delivery ratio. Both fuzzy linguistic variables are used to define the linguistic variable as four different categories. At the beginning, the residual energy of all nodes is the same. But when the nodes start to deploy and data transmission occurs, then it will be changed based on the operation. Quadratic programming is used here to model the formulation as mathematical modeling for the purpose of objective function optimization and its related constraints. Its related models are depicted in equations 1 to 4, and a complete illustration of the proposed method is shown in algorithm 1.

TABLE 3.1
Membership Functions of Residual Energy

Linguistic Variable	Range
Low	0–80
Medium	50–150
High	130–250
Very High	230–300

TABLE 3.2
Membership Functions of Packet Delivery Ratio

Linguistic Variable	Range
Poor	0–200
Moderate	150–400
Good	300–800
Very Good	700–1,000

Maximize: $\quad\quad\quad\quad\quad\quad\quad Obj_1 = (x_1)^2 + (y_1)^2$

Subject to constraints: $\quad\quad\quad e_i x_1 + p_j y_1 \geq 100$ $\quad\quad\quad\quad$ (1)

Where i = 1, 2, 3 for three different values of "low" energy, and j = 1, 2, 3 for three different values of "poor" packet delivery ratio, and the range of low energy is 0 to 80 and the range of poor packet delivery ratio is 0 to 200. In this scenario, the total number of nodes is 100.

Maximize: $\quad\quad\quad\quad\quad\quad\quad Obj_2 = (x_1)^2 + (y_1)^2$

Subject to constraints: $\quad\quad\quad e_i x_1 + p_j y_1 \geq 200$ $\quad\quad\quad\quad$ (2)

Where i = 1, 2, 3 for three different values of "medium" energy, and j = 1, 2, 3 for three different values of "moderate" packet delivery ratio, and the range of medium energy is 50 to 150 and the range of moderate packet delivery ratio is 150 to 400. In this scenario, the total number of nodes is 200.

Maximize: $\quad\quad\quad\quad\quad\quad\quad Obj_3 = (x_1)^2 + (y_1)^2$

Subject to constraints: $\quad\quad\quad e_i x_1 + p_j y_1 \geq 300$ $\quad\quad\quad\quad$ (3)

Where i = 1, 2, 3 for three different values of "high" energy, and j = 1, 2, 3 for three different values of "good" packet delivery ratio, and the range of high energy is 130 to 250 and the range of good packet delivery ratio is 300 to 800. In this scenario, the total number of nodes is 300.

Maximize: $\text{Obj}_1 = (x_1)^2 + (y_1)^2$
Subject to constraints: $e_i x_1 + p_j y_1 \geq 400$ (4)

Where i = 1, 2, 3 for three different values of "very high" energy, and j = 1, 2, 3 for three different values of "very good" packet delivery ratio, and the range of very high energy is 230 to 300 and the range of very good packet delivery ratio is 700 to 1000. In this scenario, the total number of nodes is 400.

Algorithm 1. The Proposed Method

Input: Network model, residual energy, packet delivery ratio
Output: Network lifetime
Step 1: Start.
Step 2: Initialize the network environment with parameters.
Step 3: Consider the two network parameters, i.e., residual energy and packet delivery ratio.
Step 4: Assume crisp values of energy as 300 unit and packet delivery ratio as 1,000 unit.
Step 5: No_of_nodes = 100.
Step 6: FVE=1 /* 1 for Low, 2 for Medium, 3 for High, 4 for Very High */.
Step 7: FVP=1 /* 1 for Poor, 2 for Moderate, 3 for Good, 4 for Very Good */.
Step 8: Apply fuzzification method over two input parameters.
Step 9: Assign linguistic values to the fuzzification values of input parameters.
Step 10: Decide the total number of nodes for each round as 100, 200, 300, and 400.
Step 11: Decide the decision variables as x1 and y1 for input parameters.
Step 12: Design objective function and its constraints.
Step 13: Map the constraints with linguistic variables.
Step 14: Set the second-order polynomial to the objective function.
Step 15: Run the model by using No_of_nodes, FVE, and FVP.
Step 16: Analyze the outcome of the models based on values of decision variables and objective function.
Step 17: If the outcome decreases based on number of nodes, then go to step 18. Else, go to step 13.
Step 18: Store the result.
Step 19: No_of_nodes = No_of_nodes+100.
Step 20: FVE=FVE+1.
Step 21: FVP=FVP+1.
Step 22: If No_of_nodes $\leq$ 4, then go to step 13. Else, go to step 23.
Step 23: Stop.

In equations 1 to 4, x_1 is the decision variable of residual energy for controlling the objective function of the residual energy, and y_1 is the decision variable of packet delivery ratio for controlling the objective function of packet delivery ratio. Both variables are known as input parameters of the network, and the combination of both helps derive the output parameter of the model, i.e., network lifetime, known as Obj_1, Obj_2, Obj_3, and Obj_4, for mapping with four linguistic variables as Obj_1 with "low" residual energy and "poor" packet delivery ratio, Obj_2 with "medium" residual energy and "moderate" packet delivery ratio, Obj_3 with "high" residual energy and "good" packet delivery ratio, and Obj_4 with "very high" residual energy and "very good" packet delivery ratio. In these mathematical modeling, e_i and p_j vary for $i = 1, 2, 3$ and $j = 1, 2, 3$ for different values of each iteration, where iteration is mapped based on four linguistic variables. In each iteration, the total number of nodes is changed by 100. In the first iteration, the total number of nodes is 100; in the second iteration, the total number of nodes is 200; in the third iteration, the total number of nodes is 300; and in the fourth iteration, the total number of nodes is 400.

3.4 SIMULATION AND ANALYSIS

This section deals with several illustrations of performance evaluation of mobile nodes. It is formulated in the LINGO software with nonlinear optimization technique, i.e., quadratic programming. Several simulation parameters are shown in this work, summarized in Table 3.3. The platform used in this formulation is

TABLE 3.3
Simulation Parameters

Parameter	Description
Total optimization models	4
Minimum number of nodes	100
Maximum number of nodes	400
Windows	Windows 11
Nonlinear model	4
Optimization software	LINGO
MS Office	2016
Nature of the objectives	Nonlinear
Total constraints	4×3
Nature of the constraints	Nonlinear
Total objective functions	4
Total input parameters	2
Total output parameter	1
Name of the input parameters	"Residual energy," "packet delivery ratio"
Output parameter	NetworkILifetime
Total linguistic variables	8
Linguistic variable for first input parameter	4
Name of the linguistic variable for first input parameter	Low, medium, high, very high
Linguistic variable for second input parameter	4
Name of the linguistic variable for second input parameter	Poor, moderate, good, very good

Windows 11 with MS Office 2016. Quadratic programming is used here to formulate a nonlinear objective function with constraints. The constrains used here are residual energy and packet delivery ratio. The total iteration made is 4, based on difference of 100 nodes, where the initial takes 100 nodes and at last takes 400 nodes. The total constraints is twelve, where each objective uses three constraints. The total input parameters are 2, named "residual energy" and "packet delivery ratio," and the output parameter is 1, named "network lifetime." The total linguistic variable is 8 for both input parameters. The names of the linguistic variables are "low," "medium," "high," and "very high" for the first input parameter, i.e., "residual energy," and the names of the linguistic variables for the second input parameter are "poor," "moderate," "good," and "very good."

Figures 3.1 to 3.4 show illustrations of the network lifetime of the MANET based on two input parameters, i.e., "residual energy" and "packet delivery ratio." The combination of both input parameters helps design network lifetime by optimizing three different values for each linguistic variable based on different

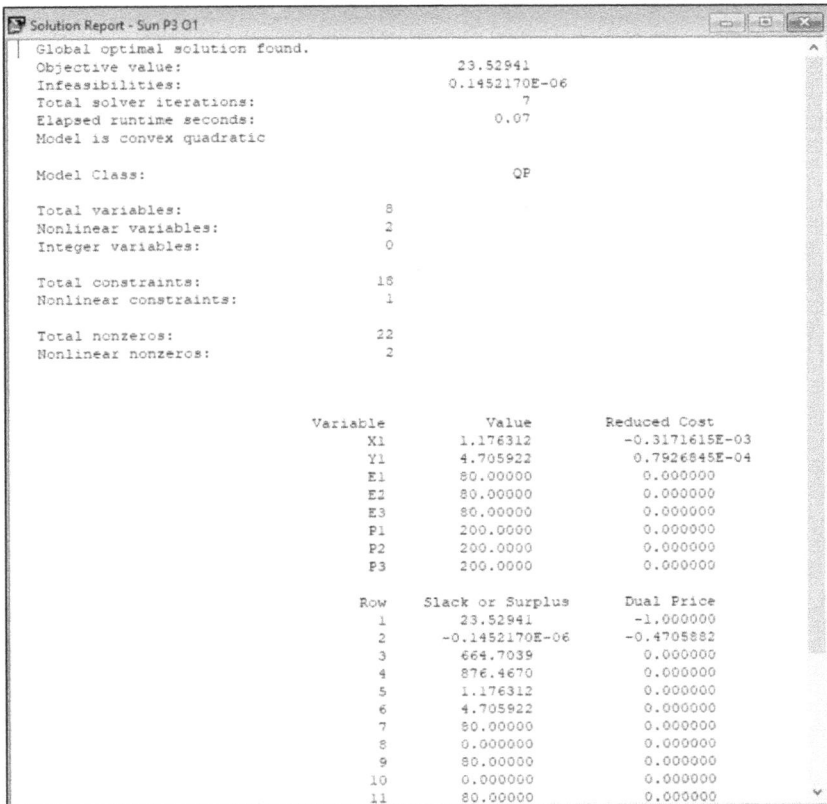

```
Solution Report - Sun P3 O1
Global optimal solution found.
Objective value:                      23.52941
Infeasibilities:                    0.1452170E-06
Total solver iterations:                 7
Elapsed runtime seconds:               0.07
Model is convex quadratic

Model Class:                            QP

Total variables:            8
Nonlinear variables:        2
Integer variables:          0

Total constraints:         16
Nonlinear constraints:      1

Total nonzeros:            22
Nonlinear nonzeros:         2


        Variable         Value        Reduced Cost
            X1          1.176312      -0.3171615E-03
            Y1          4.705922       0.7926845E-04
            E1         80.00000        0.000000
            E2         80.00000        0.000000
            E3         80.00000        0.000000
            P1         200.0000        0.000000
            P2         200.0000        0.000000
            P3         200.0000        0.000000

             Row    Slack or Surplus    Dual Price
              1        23.52941         -1.000000
              2      -0.1452170E-06     -0.4705882
              3        664.7039          0.000000
              4        876.4670          0.000000
              5        1.176312          0.000000
              6        4.705922          0.000000
              7        80.00000          0.000000
              8        0.000000          0.000000
              9        80.00000          0.000000
             10        0.000000          0.000000
             11        80.00000          0.000000
```

FIGURE 3.1 Network lifetime when number of nodes is 100.

FIGURE 3.2   Network lifetime when number of nodes is 200.

FIGURE 3.3   Network lifetime when number of nodes is 300.

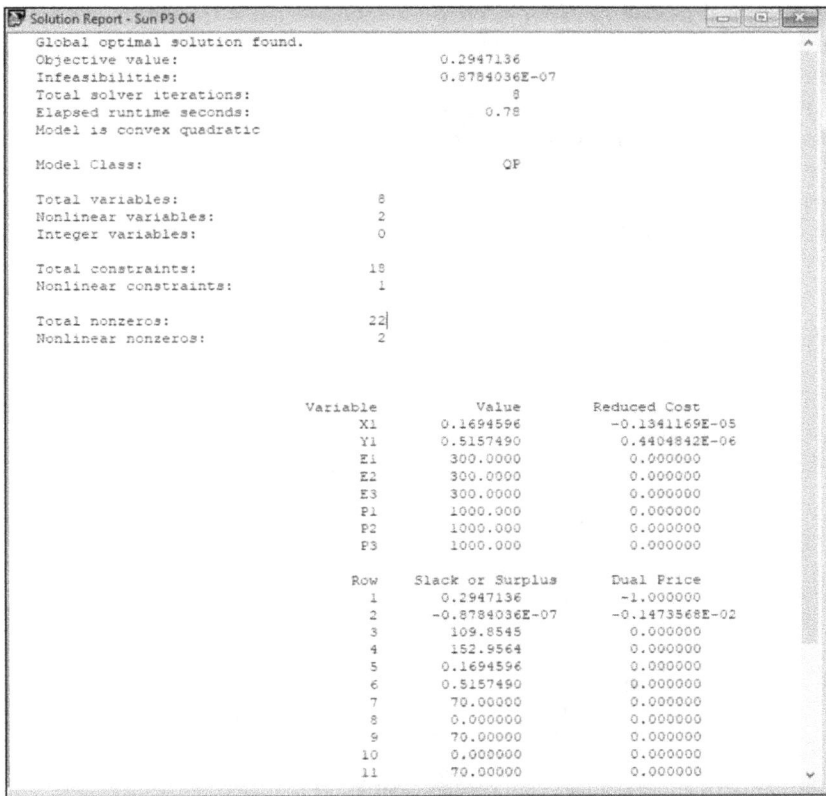

```
Solution Report - Sun P3 O4

Global optimal solution found.
Objective value:                              0.2947136
Infeasibilities:                              0.8784036E-07
Total solver iterations:                              8
Elapsed runtime seconds:                              0.78
Model is convex quadratic

Model Class:                                  QP

Total variables:                    8
Nonlinear variables:                2
Integer variables:                  0

Total constraints:                  18
Nonlinear constraints:              1

Total nonzeros:                     22
Nonlinear nonzeros:                 2


            Variable           Value         Reduced Cost
                  X1        0.1694596        -0.1341169E-05
                  Y1        0.5157490         0.4404842E-06
                  E1        300.0000          0.000000
                  E2        300.0000          0.000000
                  E3        300.0000          0.000000
                  P1        1000.000          0.000000
                  P2        1000.000          0.000000
                  P3        1000.000          0.000000

                 Row    Slack or Surplus      Dual Price
                   1        0.2947136         -1.000000
                   2       -0.8784036E-07     -0.1473568E-02
                   3        109.8545          0.000000
                   4        152.9564          0.000000
                   5        0.1694596         0.000000
                   6        0.5157490         0.000000
                   7        70.00000          0.000000
                   8        0.000000          0.000000
                   9        70.00000          0.000000
                  10        0.000000          0.000000
                  11        70.00000          0.000000
```

FIGURE 3.4 Network lifetime when number of nodes is 400.

mobile nodes. The total iteration used in this model is 4. In the first model, the total number of nodes is 100, and the network lifetime generated is 23.52941. In the second model, the total number of nodes is 200, and network lifetime generated is 0.9876543. In the third model, the total number of nodes is 300, and the network lifetime generated is 0.8419083. In the fourth model, the total number of nodes is 400, and the network lifetime generated is 0.2947136. Finally, it is observed that when the number of nodes increases, then network lifetime is decreased, based on the increasing order of linguistic behavior.

3.5 CONCLUSION

The proposed method is based on quadratic programming with the fusion of fuzzy logic to control the variation and mobility of the network. In this method, network lifetime is optimized based on two input parameters, namely, residual energy and packet delivery ratio, in four iterations. Each iteration contains one mathematical modeling with objective function and some related linguistic constraints. The linguistic constraints help optimize network lifetime based on the increasing behavior of the mobile nodes. In each iteration, the mobile nodes increase by 100, and it is observed that when the number of nodes increases, then network lifetime is decreased. Although, in this scenario, the network lifetime decreases, the linguistic behavior of the mobile nodes increases. So finally, it is concluded that the fusion of quadratic programming and fuzzy logic efficiently helps reduce imprecise input parameters of the network with the help of fuzzy triangular membership functions.

3.6 REFERENCES

[1] Prasad, R. (2021). Secure intrusion detection system routing protocol for mobile ad-hoc network. *Global Transitions Proceedings*, https://doi.org/10.1016/j. gltp.2021.10.003.

[2] Singh, N. C., & Sharma, A. (2020). Resilience of mobile ad hoc networks to security attacks and optimization of routing process. *Materials Today: Proceedings*, https:// doi.org/10.1016/j.matpr.2020.09.622.

[3] Anand, M., Balaji, N., Bharathiraja, N., & Antonidoss, A. (2021). A controlled framework for reliable multicast routing protocol in mobile ad hoc network. *Materials Today: Proceedings*, https://doi.org/10.1016/j.matpr.2020.10.902.

[4] Patsariya, M., & Rajavat, A. (2021). Node capability-based route selection on mobile ad hoc network. *Materials Today: Proceedings*, https://doi.org/10.1016/j. matpr.2020.12.210.

[5] Kumar, B. V. S., & Padmavathy, N. (2020). A hybrid link reliability model for estimating path reliability of mobile ad hoc network. *Procedia Computer Science*, 171, 2177–2185.

[6] Prasad, R. (2021). Enhanced energy efficient secure routing protocol for mobile ad-hoc network. *Global Transitions Proceedings*, https://doi.org/10.1016/j.gltp.2021.10.001.

[7] Singh, N. C., & Sharma, A. (2020). Resilience of mobile ad hoc networks to security attacks and optimization of routing process. *Materials Today: Proceedings*, https:// doi.org/10.1016/j.matpr.2020.09.622.

[8] Das, S. K. (2021). Smart design and its applications: Challenges and techniques. *Nature-Inspired Computing for Smart Application Design*, 1.

[9] Dey, N., Hassanien, A. E., Bhatt, C., Ashour, A., & Satapathy, S. C. (Eds.). (2018). *Internet of Things and Big Data Analytics toward Next-generation Intelligence* (Vol. 35). Springer, Berlin.

[10] Singh, A. K., Pamula, R., Jain, P. K., & Srivastava, G. (2021). An efficient vehicular-relay selection scheme for vehicular communication. *Soft Computing*, 1–17.

[11] Das, S. K., Maheswari, V., & Sharma, A. (2021). Wireless networks: Applications, challenges, and security issues. In *Architectural Wireless Networks Solutions and Security Issues* (pp. 1–10). Springer, Singapore.

[12] Anand, M., Balaji, N., Bharathiraja, N., & Antonidoss, A. (2021). A controlled framework for reliable multicast routing protocol in mobile ad hoc network. *Materials Today: Proceedings*, https://doi.org/10.1016/j.matpr.2020.10.902.

[13] Tahir, A., Shah, N., Abid, S. A., Khan, W. Z., Bashir, A. K., & Zikria, Y. B. (2021). A three-dimensional clustered peer-to-peer overlay protocol for mobile ad hoc networks. *Computers & Electrical Engineering*, 94, 107364, https://doi.org/10.1016/j. compeleceng.2021.107364.

[14] Singh, A. K., & Pamula, R. (2021). An efficient and intelligent routing strategy for vehicular delay tolerant networks. *Wireless Networks*, 27(1), 383–400.

[15] Das, S. K., Samanta, S., Dey, N., & Kumar, R. (Eds.). (2020). *Design Frameworks for Wireless Networks*. Springer, Singapore.

[16] Sharma, A. S., & Kim, D. S. (2021). Energy efficient multipath ant colony based routing algorithm for mobile ad hoc networks. *Ad Hoc Networks*, 113, 102396, https://doi.org/10.1016/j.adhoc.2020.102396.

[17] Lee, S., Youn, J., & Jung, B. C. (2020). A cooperative phase-steering technique in spectrum sharing-based military mobile ad hoc networks. *ICT Express*, 6(2), 83–86.

[18] Singh, A. K., & Pamula, R. (2021). Vehicular delay tolerant network based communication using machine learning classifiers. *Architectural Wireless Networks Solutions and Security Issues*, 195, https://doi.org/10.1007/978-981-16-0386-0_11

[19] Das, S. K., Samanta, S., Dey, N., Patel, B. S., & Hassanien, A. E. (Eds.). (2021). *Architectural Wireless Networks Solutions and Security Issues*. Springer, Singapore.

[20] Keerthika, M., & Shanmugapriya, D. (2021). Wireless sensor networks: Active and passive attacks-vulnerabilities and countermeasures. *Global Transitions Proceedings*, 2(2), 362–367.

[21] Wan, J., & Chen, J. (2022). AHP based relay selection strategy for energy harvesting wireless sensor networks. *Future Generation Computer Systems*, 128, 36–44.

[22] Singh, A. K., Pamula, R., & Srivastava, G. (2022). An adaptive energy aware DTN-based communication layer for cyber-physical systems. *Sustainable Computing: Informatics and Systems*, 100657, https://doi.org/10.1016/j.suscom.2022.100657.

[23] De, D., Mukherjee, A., Das, S. K., & Dey, N. (Eds.). (2020). *Nature Inspired Computing for Wireless Sensor Networks*. Springer, Singapore.

[24] Shaban, W. M., Rabie, A. H., Saleh, A. I., & Abo-Elsoud, M. A. (2021). Detecting COVID-19 patients based on fuzzy inference engine and Deep Neural Network. *Applied Soft Computing*, 99, 106906, https://doi.org/10.1016/j.asoc.2020.106906.

[25] Zahra, S. R., Chishti, M. A., Baba, A. I., & Wu, F. (2021). Detecting COVID-19 chaos driven phishing/malicious URL attacks by a fuzzy logic and data mining based intelligence system. *Egyptian Informatics Journal*, https://doi.org/10.1016/j.eij.2021.12.003.

[26] Ahmed, H. I., Nasr, A. A., Abdel-Mageid, S. M., & Aslan, H. K. (2021). DADEM: Distributed attack detection model based on big data analytics for the enhancement of the security of internet of things (IoT). *International Journal of Ambient Computing and Intelligence (IJACI)*, 12(1), 114–139.

[27] Sholla, S., Mir, R. N., & Chishti, M. A. (2021). A fuzzy logic-based method for incorporating ethics in the internet of things. *International Journal of Ambient Computing and Intelligence (IJACI)*, 12(3), 98–122.

[28] Balusa, B. C., & Gorai, A. K. (2021). Development of fuzzy pattern recognition model for underground metal mining method selection. *International Journal of Ambient Computing and Intelligence (IJACI)*, 12(4), 64–78.

Section 2

Modelling and Aggregation

Modelling and Aggregation

4 Game Theory– Based Conflicting Strategy Management Technique in Wireless Sensor Network

Santosh Kumar Das, Aman Kumar Tiwari, Somnath Rath, and Joydev Ghosh

CONTENTS

4.1 Introduction 57
4.2 Related Works.. 59
4.3 Preliminaries 63
    4.3.1 Game Theory... 63
    4.3.2 Fuzzy Logic 64
4.4 Proposed Method... 64
4.5 Simulation and Analysis 68
4.6 Conclusions... 72
4.7 References.. 72

4.1 INTRODUCTION

The application of the wireless sensor network, or WSN, increases rapidly due to its dynamic and fusion characteristics [1–3]. It consists of tiny nodes that are known as sensor nodes that help sense several types of information. It helps model several applications based on services. Each node is autonomous, based on several connections of internet services. Several components with WSN are shown in Figure 4.1 for the purpose of sensing information. The main issue of this intelligent network is the battery of the nodes because each node consists of a limited-capacity battery for sensing information [4]. In this figure, three types of nodes are defined as low-energy enable node, medium-energy enable node, and high-energy enable node. These three types of nodes are covered with radio range, where some cross points indicate combination of two types of nodes. Each region is directly connected with the BS for communication in terms of data and information. BS is connected with the internet, and the internet with the user, for completing the actual communication. The internet provides the services to the BS. The user that's connected with the BS and the internet analyzes the actual information from time to time and stores it for future processing.

Details of the operation in Figure 4.1 are shown in Figure 4.2. It shows the working principle of the WSN based on different types of sensor nodes. It contains three linguistic regions that are connected to one another to achieve the purpose of the goal of the WSN. The overall region is connected with the internet for processing data and information. In this mode, the combined sensor nodes first

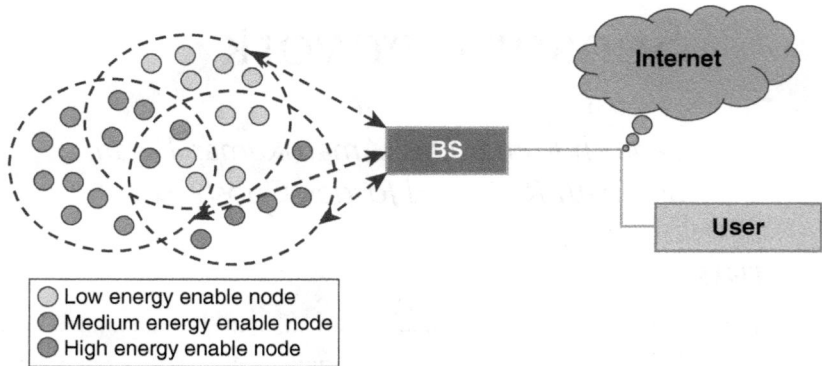

FIGURE 4.1 Wireless sensor networks.

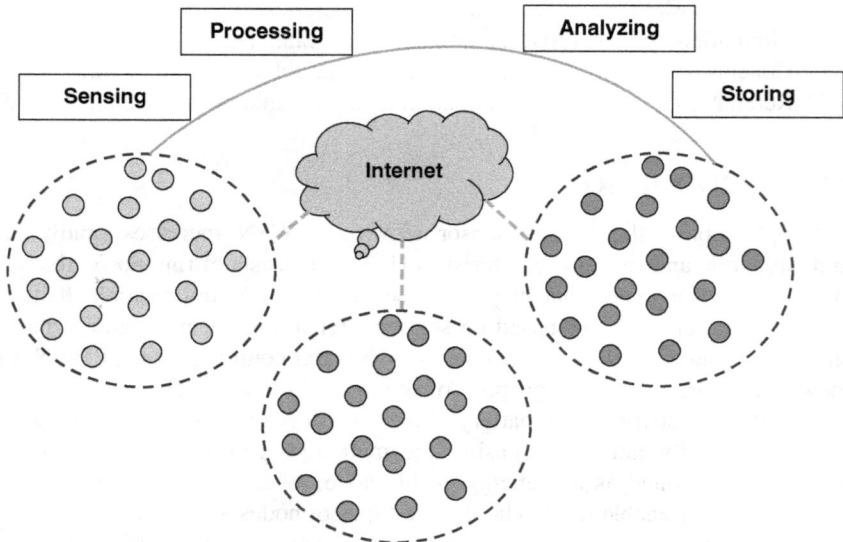

FIGURE 4.2 Working principle of WSN based on different nodes.

sense the environmental information, then process it for reducing interference or extracting actual information, then analyze it for required information based on user query, and finally store it for future references or for future purposes.

There are several applications of the WSN in terms of different services, but it has one crucial limitation—that the sensor node consists of a small battery, and this battery contains low energy. This low energy is insufficient during any mission or operation. Due to this limited energy capacity, transaction fails continuously and simultaneously. It causes several types of uncertainty and interference in the network that creates a critical chaos. So it fails to decrease the performance of each network metrics. It also helps decrease network lifetime. So in this paper, an intelligent protocol is designed to estimate the nonlinear parameters and fluctuated information of the network. The main intelligent technique chosen in this work is game theory. Game theory is an applied mathematical optimization tool or technique that helps optimize network parameters. It helps map the strategy of the network to model the actual percentage of parameter selection. So that path of the network does not fail.

The remainder of this article, divided into sections, discusses the details of existing work based on network optimization so that it will guide the proposed method. The next section discusses technique background information based on the proposed method. It takes a crucial part of the proposed method for the purpose of formulation. The next section discusses the detailed illustration of the proposed method to achieve the purpose of the goal. The next section discusses simulation of the method. The last section concludes the paper.

4.2 RELATED WORKS

The related works section describes several works for the purpose of network formulation and its optimization based on several parameters. These parameters also constraint network management elements. So in this section, several works have been proposed to enhance network lifetime and its formulation. S. K. Das [5] proposed a method for the purpose of the application of the design system and its management. It helps in several application management systems that deal with several challenges and issue management. It helps model several information systems based on smart application systems. It helps model several security management systems based on emergency management and application systems. It helps give new guidelines that will help model several issues in terms of solution. This solution helps model several services based on real-life application management systems. It helps in several monitoring and application management systems based on emergency and security modeling systems that are, in turn, based on services. N. Dey [6] proposed a method for a nature-inspired optimization system that helps in modeling several variants of information. The work is based on several application and modeling systems that help model several information systems. There are several works and applications combined in this application system that help in modeling nonlinear optimization efficiently and effectively. The contribution of this book helps model several optimizations with the context of real-life application. It helps in modeling several sources of information based on a multiapplication system on power management. P. N. Mahalle et al. [7] illustrated an underwater system for wireless sensor network that helps model several applications. The work of this book helps model several sensor applications for modeling a dynamic information system. It helps model several static and dynamic information systems based on the nature of the sensor information system. Several architectural information systems are modeled for a dynamic application system. It helps in modeling several flows of information based on types of component analysis that help in component analysis. Keerthika and Shanmugapriya [8] designed a method that combines a passive and active attack for the purpose of illustration. This illustration is based on some countermeasures system that helps in vulnerabilities systems. It helps model the application based on environmental analysis, which helps deal with a protection system based on commercial analysis. The communication of the system is based on the deployment of some challenges along with some issues. It helps in defensive analysis based on vulnerability analysis of some factors of information. Singh et al. [9] proposed an adaptive method for the purpose of creating an energy-aware system that is used for communication purposes. This method is used for physical communication systems based on a delay-tolerant system. It helps in modeling applications based on a cyber-physical system that is used for wireless sensor networks. It helps model applications based on delay-tolerant and prediction systems. The application is based on a mobile application for operating several operations. The work is simulated based on the network environment for predicting network lifetime. Das et al. [10] designed an application management illustration for the purpose of wireless network systems and services. It helps model several applications based on different security and challenge system management. It helps in issues on management based on several parts of the communication system. It helps adopt several solutions based on some higher analysis and management. The work is based on complexity management, which helps model several solutions within the context of management based on different variations of a wireless network. Wan and Chen [11] designed a strategy for energy analysis and mechanism for harvesting analysis. The work is based on a WSN purpose of modeling. It helps model several cooperative analyses for node analysis. It defines some probability based on relay node detection. The main purpose of this analysis to solve network performance based on certain factors. It helps model the application and save the actual energy. It uses mathematical modeling for analyzing data and its parameters. It helps enhance energy based on a solar energy system and its cooperation. Singh et al. [12] designed an efficient modeling method for the purpose of helping for vehicular communication. The work is based on communication purposes and modeling. It helps several applications model several vehicular relay managements for the purpose of ensuring efficient network communication. The work is based on a utilization purpose that helps model several potential system managements. It helps to model several relay application and network modeling. It helps predict several vehicular communications based on selfish note prediction. It helps in strategy management for tracking network performance. Misra et al. [13]

designed an implementation method based on the fusion of FPGA and NLOS. The work is based on distance analysis and its estimation system. It helps in modeling several applications that help elderly modeling. The work is designed for the purpose of an indoor system that helps in WSN. It helps in location analysis based on the ZigBee network. The work uses a programmable gate array system and its modeling. This modeling uses artificial neural network to estimate different errors and improve network lifetime. It uses a hybridization method for modeling several complexities based on suitable analysis. Singh and Pamula [14] designed a method for the purpose of intelligent communication and routing model. The work is based on a vehicular communication system based on strategy management. It helps based on strategy behavior modeling and analysis, which helps enhance network lifetime. The application is deployed in several purposes based on protocol management. It helps track analysis in novel behavior based on delay-tolerant management. It helps utilize and analyze the system based on vehicular communication. It helps outperform the result based on several variations. De et al. [15] proposed an illustration for the purpose of challenges and application management services. This service is based on the application of a wireless sensor network system and its variations. It helps model several challenges and application management systems. It helps deal with several algorithms of wireless networks based on variations and their analysis. It helps guide several working principles and information systems based on service management. It helps deal with algorithm analysis that helps manage several applications of the system. Wang and Hu [16] designed a hole-detection method for handling several issues based on WSN. The network is based on a clustering method and algorithm that uses some gap coverage analysis. It helps to analyze multihop management systems for rational deployment. It helps distance parameter systems and vulnerability detection, which helps in coverage modeling and its parameters. It overcomes the limitation of several determination systems for edge node modeling. It helps determine random walk connection and its management. Singh and Pamula [17] designed a delay management system for the purpose of a communications-based system. The work is based on a machine learning–based application that helps in several purposes. The work is based on several classifier models of analysis based on the solution for vehicular networks. It helps deal with several network metrics based on certain parameters of the network. The work is based on strategy management for handling the inadequate design of the system. It helps outperform the result based on the consideration of system modeling. Temene et al. [18] illustrated a survey based on mobility analysis and prediction for WSN. The work is based on IoT and WSN both for detailed illustration. It helps to model several mobile nodes. There are several mobile nodes that play different roles, such as the sink node, mobile node, source node, etc. The combination of all nodes helps model several congestions and its related mitigations. It helps in the predecessor analysis of IoT, which helps in several directions. The work helps model several evaluations based on different algorithms. Singh et al. [19] designed a method for the purpose of strategy management that is based on delay analysis. It helps maintain several network management systems based on some replication analyses.

It helps deal with and manage several packet control management systems. It helps manage several data delivery and resource management systems for the purpose of communication systems handling. The work is based on routing analysis that helps deal with and manage intermediate node analysis systems. The work deals with several assists for the purpose of replication management. This replication is based on a control system and a parameter-dealing system. Das et al. [20] designed a book for the purpose of generating an architectural solution system based on wireless network. This service is based not only on network but also on several systems and information based on the architecture of the network. It helps in modifications based on the architecture of the system. It helps model several issues of the network. Several issues are used and deal with the system. Some of the issues mentioned are energy efficiency system, network lifetime system, resource management, data aggregation system, etc. It helps model several solutions and in security management of the wireless network. Yousefpoor et al. [21] designed a secure method for WSN as a review paper that helps model several issues in the network. The work is based on a data aggregation method that helps reduce the attack in the system. It helps in countermeasures for several issues with the context of attack measurement. This review is also based on the industrial internet of things system and its modeling. It helps manage several issues with the context of solution measurement. It helps save energy and increase security of the system based on an authentication system. Zhang and Mao [22] designed a multifactor system for authentic purposes. The work is based on a protocol system that helps model the application. It helps model several recognitions to exercise the physical system. It is based on the ZigBee network system, which helps model several scope identifications. It helps in security analysis and the recognition of several applications based on component analysis and its management. It helps model several information based on radio frequency analysis. It helps design the system based on security analysis for connection management of the network. Huanan et al. [23] designed a security-based application system for the purpose of wireless sensor network. It helps model several systems for handling several intrusion and detection systems. It helps model several systems based on the foundation of the network. It helps in the study of the system for reducing several threats. It helps in modeling some analysis and emphasizes communication and modeling. It helps in handling several security systems for designing some analysis and its modeling. Das et al. [24] designed a book for the purpose of creating a smart application design. The work of this book is based on smart application with the fusion of smart computing. It helps in modeling several applications based on some service and management. It helps model several nature-inspired applications based on computing application services. It helps model and give new insight in the subarea of network modeling, data analysis and prediction, network lifetime management, resource and energy management, etc. It helps model several information systems for the management of dynamic applications and planning and services. Shaban et al. [25] designed a patient analysis system used for inference system and design. It is based on several information that help model several neural network systems and models. The work is based on fuzzy inference modeling system that helps model some issues of COVID-19. The issue is based on deep learning and fuzzy system modeling. It helps model several network information based on strategy management. It helps validate several cross analysis and validation that helps in accuracy modeling. It helps model several detections for coronavirus analysis for helping and modeling based on a prevention system. Zahra et al. [26] designed a method for a data-driven system that is based on intelligence system maintenance and information system. The work is based on URL system information and modeling that helps analyze and design several malicious and phishing information system. The work is completely based on fuzzy logic modeling. It helps deal with several uncertainty management for handling pandemic analysis. It helps model several control and unprecedented information systems. The work helps model several cybercriminal information analyses to deal with several ransomware information system for impact analysis. Das et al. [27] designed a system and model for the purpose of wireless network and application. It helps model several information based on service management and application modeling. The work is based on several information and management systems, such as energy resource management and modeling. It helps deal with several security and privacy management systems that help in design and its enhancement. The work is based on troubleshooting and automation system for network lifetime management and its enhancement based on security, modeling, and learning [28–30]. It gives several protocols for the purpose of design perspective and modeling.

4.3 PRELIMINARIES

A short description of the technique background is discussed in this section to help understand the goal.

4.3.1 GAME THEORY

Game theory is an optimization technique which is used as applied mathematical model. In this model, the number of entities is two or more than two, depending on the situation. Each entity is known as a player. This game helps manage interaction among players. This technique is increasingly used in several areas, such as political science, economics, sociology, biology, engineering, management, etc. Equation 4.1 shows the basic elements of the game theory.

$$G = \{P, A, U\} \qquad (4.1)$$

Where P is the set of players, A is the set of actions of the players, and U is the set of utility functions. *Utility function* indicates that one player wants to maximize its own profit, then its opponent wants to minimize its losses.

In game theory, using payoff or utility function, a player can represent its own objective. This objective is mapped into an objective function along with constraints. Both players have separate objective functions along with constraints that deal with the proposed game theory. Nash equilibrium is a mathematical modeling under game theory technique which is used in modeling interaction of the players efficiently. In this modeling, one rule is established between players, that no one deviates from the rules, so that no player can get more profit or no opponent can get more loss in playing the game.

4.3.2 FUZZY LOGIC

It is a multivalued logic which paves the gap of traditional logic or crisp logic. It deals with imprecise information and is used to control noise and uncertainty of the system. Its mathematical model is shown in Equation 4.2.

$$FL = \{x, \mu_A(x)\} \tag{4.2}$$

Where x is the element and $\mu_A(x)$ is the membership value of the x.

The element of the fuzzy logic is a crisp value which be anything, be it an integer or a fraction, depending on the problem and its parameters. Membership value is derived from membership function, such as triangular membership function, trapezoidal membership function, sigmoid membership function, etc. Its mathematical model is represented by $\mu_A(x)$, where A is the fuzzy set, x is an element of the fuzzy set, and $\mu_A(x)$ is the membership value of the element x, which has range between 0 and 1.

4.4 PROPOSED METHOD

In this section, the proposed method is illustrated briefly. It is a fusion of game theory and fuzzy logic where a noncooperative game theory technique is used for handling the dynamic topology of the WSN. Noncooperative game theory is the opposite of cooperative game theory, where players do not compromise or agree on the situation for handling the strategy of the game theory for resolving the solution of the game. The proposed game model is shown in equation 4.3, where N is the set of nodes that act as players and L is the set of links or edges that carry different strategies. An environment of the network is shown in Table 4.1 that consists of Node_id, which indicates identification number of the nodes, and energy, which indicates residual energy of the node, shown in equation 4.4, which is the difference between initial energy and energy consumed. *Energy consumed*, shown in equation 4.5, is the combination values of energy used in sending packet and in receiving packet. *Mobility* indicates movement of the dynamic sensor

TABLE 4.1
Environment of the Sensor Nodes

Node_id	Energy	Mobility	Control Packet	Bandwidth

nodes, while *control packet* indicates total amount of data packet sent or received by a particular node. *Bandwidth* indicates speed of sending and receiving data packet of sensor nodes.

$$Game = G(N, L) \tag{4.3}$$

$$Residual\_Energy = Initial\_Energy-Energy\_Consumed \tag{4.4}$$

$$Energy\_Consumed = Energy\_SndPkt + Energy\_RcvPkt \tag{4.5}$$

Details of the node environment are shown in Table 4.2. In this environment, energy is considered as 100 J, mobility is considered as 200 bits/sec, control packet is considered as 300 unit, and bandwidth is considered as 400 unit. All these parameters are generated randomly in the network environment. Here, one threshold value is considered based on equation 4.6 for deciding two players as player A and player B. The nodes, having energy level greater than or equal to the considered threshold value, are under player A, and the nodes having energy level less than the considered threshold value are under player B. Player A is considered as team A, and player B is considered as team B, as shown in Tables 3 and 4. In these tables, the total number of nodes in team A is t1, and the total number of nodes in team B is t2, where t1 and t2 may or may not be the same, but the combination of both t1 and t2 is t.

$$Energy\_th = mean\ (Energy_{low}, Energy_{high}) \tag{4.6}$$

After deciding both teams as team A and team B for both players, apply fuzzy logic for fuzzifying input data for reducing uncertainty from the network. In this mode, triangular fuzzy membership function is used for fuzzification purpose. Figure 4.3 shows membership function of fuzzy logic. In this figure, the x-axis indicates universe of discourse (UOD), and the y-axis indicates degree of membership (DOM). The combination of both helps produce membership value. In the proposed method, there are four input parameters, which are energy or residual

TABLE 4.2
Details of the Environment

Node ID	Energy	Mobility	Control Packet	Bandwidth
n_1	e_1	m_1	c_1	b_1
n_2	e_2	m_2	c_2	b_2
...	...	...	...	...
...	...	...	...	...
n_{t-1}	e_{t-1}	m_{t-1}	c_{t-1}	b_{t-1}
n_t	e_t	m_t	c_t	b_t

energy, mobility, control packet, and bandwidth. All considered, values of the input parameters are crisp values that are mapped into the x-axis. In this membership function, apply randomness value as "x" for all input parameters and find their membership values. After finding the membership values, the payoff matrix is calculated based on equation 4.7, and the payoff matrix is shown in Table 4.5. In Table 4.5, each cell value is known as the payoff value of the game, which has a range between 0 and 1, known as the fuzzy value.

$$\text{Payoff\_value} = \text{Mean}(\mu_{\text{TeamA}}(x), \mu_{\text{TeamB}}(x)) \qquad (4.7)$$

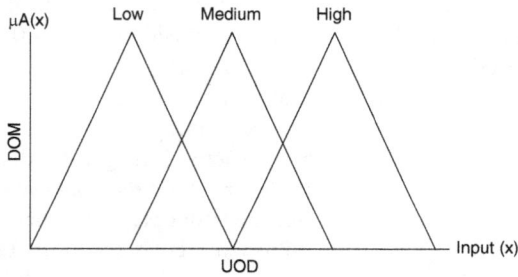

FIGURE 4.3   Membership function of fuzzy logic.

TABLE 4.3
Details of Player A or Team A

Node ID	Energy	Mobility	Control Packet	Bandwidth
n_1	e_1	m_1	c_1	b_1
n_2	e_2	m_2	c_2	b_2
...	...	...	...	...
...	...	...	...	...
n_{t1-1}	e_{t1-1}	m_{t1-1}	c_{t1-1}	b_{t1-1}
n_{t1}	e_{t1}	m_{t1}	c_{t1}	b_{t1}

TABLE 4.4
Details of Player B or Team B

Node ID	Energy	Mobility	Control Packet	Bandwidth
n_1	e_1	m_1	c_1	b_1
n_2	e_2	m_2	c_2	b_2
...	...	...	...	...
...	...	...	...	...
n_{t2-1}	e_{t2-1}	m_{t2-1}	c_{t2-1}	b_{t2-1}
n_{t2}	e_{t2}	m_{t2}	c_{t2}	b_{t2}

TABLE 4.5
Payoff Matrix of the Proposed Game Model

		Team B			
		E	M	C	B
Team A	E	[0, 1]	[0, 1]	[0, 1]	[0, 1]
	M	[0, 1]	[0, 1]	[0, 1]	[0, 1]
	C	[0, 1]	[0, 1]	[0, 1]	[0, 1]
	B	[0, 1]	[0, 1]	[0, 1]	[0, 1]

The network topology of the WSN is dynamic, and it consists of several types of variations, so in this payoff matrix, there is no saddle point. A saddle point indicates where the minimax value is equivalent to the maximin value based on the row and column data. Hence, the decision maker of the network solves it with the help of linear programming. The decision maker of the WSN maps multiple player objectives based on a noncooperative game theory mechanism using linear programming by the fusion of objective functions along with constraints. A detailed method is shown in Figure 4.4. Equation 4.8 shows elements of a proposed game model, where V is the value of the game, $r1$ to $r4$ are probabilities of selecting strategy of E, M, C, and B of team A, and $s1$ to $s4$ are probabilities of selecting strategy of E, M, C, and B of team B. The proposed payoff matrix along with its probability values are shown in Table 4.6. In this model, team A is considered as the winner, and team B is considered as the loser. So team A's objective is to maximize the expected gains, so maximize the value of V and minimize the value of $\dfrac{1}{V}$. Team B's objective is to minimize the expected loss, so minimize the value of V and maximize the value of $\dfrac{1}{V}$. The proposed optimization model for team A and team B is shown in equations 4.9 and 4.10.

$$\text{Game\_model} = \{V, (r_1, r_2, r_3, r_4), (s_1, s_2, s_3, s_4)\} \tag{4.8}$$

Maximize $Z_1 = V$

Subject to: $\mu_{EE_A}(x)\, r_1 + \mu_{ME_A}(x)\, r_2 + \mu_{CE_A}(x)\, r_3 + \mu_{BE_A}(x)\, r_4 \geq V$

$\mu_{EM_A}(x)\, r_1 + \mu_{MM_A}(x)\, r_2 + \mu_{CM_A}(x)\, r_3 + \mu_{BM_A}(x)\, r_4 \geq V$

$\mu_{EC_A}(x)\, r_1 + \mu_{MC_A}(x)\, r_2 + \mu_{CC_A}(x)\, r_3 + \mu_{BC_A}(x)\, r_4 \geq V \tag{4.9}$

$\mu_{EB_A}(x)\, r_1 + \mu_{MB_A}(x)\, r_2 + \mu_{CB_A}(x)\, r_3 + \mu_{BB_A}(x)\, r_4 \geq V$

Minimize $Z_2 = V$

Subject to: $\mu_{EE_B}(x)\, s_1 + \mu_{EM_B}(x)\, s_2 + \mu_{EC_B}(x)\, s_3 + \mu_{EB_B}(x)\, s_4 \leq V$

$\mu_{ME_B}(x)\, s_1 + \mu_{MM_B}(x)\, s_2 + \mu_{MC_B}(x)\, s_3 + \mu_{MB_B}(x)\, s_4 \leq V$

$\mu_{CE_B}(x)\, s_1 + \mu_{CM_B}(x)\, s_2 + \mu_{CC_B}(x)\, s_3 + \mu_{CB_B}(x)\, s_4 \leq V \tag{4.10}$

$\mu_{BE_B}(x)\, s_1 + \mu_{BM_B}(x)\, s_2 + \mu_{BC_B}(x)\, s_3 + \mu_{BB_B}(x)\, s_4 \leq V$

TABLE 4.6

Payoff Matrix Along with Probability Values

		Team B				
		E	**M**	**C**	**B**	**Probability**
Team A	E	[0, 1]	[0, 1]	[0, 1]	[0, 1]	r_1
	M	[0, 1]	[0, 1]	[0, 1]	[0, 1]	r_2
	C	[0, 1]	[0, 1]	[0, 1]	[0, 1]	r_3
	B	[0, 1]	[0, 1]	[0, 1]	[0, 1]	r_4
	Probability	s_1	s_2	s_3	s_4	

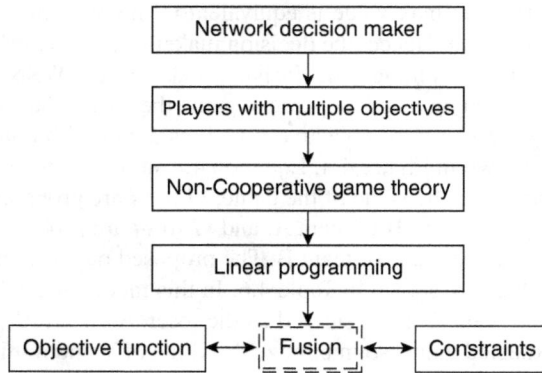

FIGURE 4.4   Decision maker process for fusion of objective function and constraints.

In the proposed method, the final values are r_1, r_2, r_3, and r_4 for team A, and s_1, s_2, s_3, and s_4 for team B. The values indicate Player A is the winner if it selects the strategies of energy, mobility, control packet, and bandwidth based on the probabilities r_1, r_2, r_3, and r_4. And Player B is the loser if it selects the strategies of energy, mobility, control packet, and bandwidth based on probabilities s_1, s_2, s_3, and s_4. Hence, the final outcomes are evaluated using linear programming.

4.5   SIMULATION AND ANALYSIS

The simulation of the proposed method is done in the Python software, which is one of the popular programming languages. The details of the parameters are

---

TABLE 4.7
Simulation Parameters

Parameter	Description
Windows	Windows 11
Programming	Python
MS Office	2016
Energy	100
Mobility	200
Control packet	300
Bandwidth	400

---

shown in Table 4.7. The parameter with the context of simulation used in this system is the Windows 11 operating system. The Office package used is MS Office 2016. The unit of parameters used in this method are 200 unit for mobility, 100 unit for energy, 300 unit for control packet, and 400 unit for bandwidth. The combination of these parameters helps optimize the network efficiently based on some variation and round. Figure 4.5 shows the network environment of the WSN, where initially the total number of nodes is entered as 50. It contains node ID, energy, mobility, control packet, and bandwidth of the sensor nodes. In this model, the user can access information on the particular sensor node, which is shown in Figure 4.6.

After creating the environment of the sensor node, assign the network parameters within the sensor nodes in which the decision maker can easily access information on the particular node. A threshold value, which is the mean value of the lowest and highest energy of the node, for each scenario is decided for creating two teams as team A and team B. It decides the players of the game as player A and player B. In this proposed model, network topology is considered as the dynamic that changes rapidly based on the scenario of the running program. Figures 4.7 and 4.8 show scenarios of both players as team A and team B, where player A consists of 23 nodes and player B consists of 27 nodes. Both scenarios have the same environment as mentioned in Figure 4.5. Figure 4.9 shows fuzzy membership function, which is used for generating crisp value to fuzzy value by using fuzzy linguistic variable. In this model, three linguistic variables are considered as "poor" for "low," "average" for "medium," and "good" for "high." It is calculated by Python library function, which is also known as tip value. Final payoff matrix for both teams is shown in Figure 4.10, which is used in linear programming for generating probabilities of selecting strategies of both players as r_1, r_2, r_3, and r_4 for team A and s_1, s_2, s_3, and s_4 for team B.

Details of ALL Players

Node Id	Energy	Mobility	Control Packet	Bandwidth
0	98	137	232	29
1	46	116	144	131
2	1	44	190	377
3	80	187	281	275
4	98	43	110	326
5	83	172	288	14
6	35	62	30	4
7	87	4	23	297
8	75	11	46	134
9	98	94	93	335
10	83	54	32	252
11	57	83	21	231
12	62	83	195	14
13	5	109	136	172
14	89	157	105	377
15	6	100	40	61
16	19	51	196	92
17	83	147	40	211
18	32	98	27	285
19	45	125	2	284
20	88	181	193	224
21	94	70	142	238
22	20	173	9	291
23	27	96	142	224
24	40	196	78	367
25	39	135	87	114
26	65	8	169	310
27	15	3	206	256
28	35	81	221	168
29	67	39	275	292
30	43	25	261	274
31	33	117	154	307
32	86	111	27	204
33	38	6	44	202
34	19	2	268	396
35	81	40	200	233
36	28	30	120	394
37	81	199	281	167
38	97	161	66	285
39	53	46	186	19
40	87	33	16	154
41	64	57	82	181
42	65	130	10	39
43	33	184	95	381
44	34	169	271	184
45	76	33	116	268
46	87	92	214	201
47	5	96	1	30
48	90	177	78	18
49	26	191	228	352

FIGURE 4.5   Network environment where nodes are from IDs 0 to 49.

```
ID 8
Energy 75
Mobility 11
ControlPacket 46
Bandwidth 134
```

FIGURE 4.6   Details of node ID 8.

```
                          TEAM A
              No of nodes in A:- 23
+---------+--------+----------+----------------+-----------+
| Node Id | Energy | Mobility | Control Packet | Bandwidth |
+---------+--------+----------+----------------+-----------+
|    1    |   46   |   116    |      144       |    131    |
|    2    |    1   |    44    |      190       |    377    |
|    6    |   35   |    62    |       30       |      4    |
|   13    |    5   |   109    |      136       |    172    |
|   15    |    6   |   100    |       40       |     61    |
|   16    |   19   |    51    |      196       |     92    |
|   18    |   32   |    98    |       27       |    285    |
|   19    |   45   |   125    |        2       |    284    |
|   22    |   20   |   173    |        9       |    291    |
|   23    |   27   |    96    |      142       |    224    |
|   24    |   40   |   196    |       78       |    367    |
|   25    |   39   |   135    |       87       |    114    |
|   27    |   15   |     3    |      206       |    256    |
|   28    |   35   |    81    |      221       |    168    |
|   30    |   43   |    25    |      261       |    274    |
|   31    |   33   |   117    |      154       |    307    |
|   33    |   38   |     6    |       44       |    202    |
|   34    |   19   |     2    |      268       |    396    |
|   36    |   28   |    30    |      120       |    394    |
|   43    |   33   |   184    |       95       |    381    |
|   44    |   34   |   169    |      271       |    184    |
|   47    |    5   |    96    |        1       |     30    |
|   49    |   26   |   191    |      228       |    352    |
+---------+--------+----------+----------------+-----------+
```

FIGURE 4.7   Detail of nodes for player A.

```
                          TEAM B
              No of nodes in B:- 27
+---------+--------+----------+----------------+-----------+
| Node Id | Energy | Mobility | Control Packet | Bandwidth |
+---------+--------+----------+----------------+-----------+
|    0    |   98   |   137    |      232       |     29    |
|    3    |   80   |   187    |      281       |    275    |
|    4    |   98   |    43    |      110       |    326    |
|    5    |   83   |   172    |      288       |     14    |
|    7    |   87   |     4    |       23       |    297    |
|    8    |   75   |    11    |       46       |    134    |
|    9    |   98   |    94    |       93       |    335    |
|   10    |   83   |    54    |       32       |    252    |
|   11    |   57   |    83    |       21       |    231    |
|   12    |   62   |    83    |      195       |     14    |
|   14    |   89   |   157    |      105       |    377    |
|   17    |   83   |   147    |       40       |    211    |
|   20    |   88   |   181    |      193       |    224    |
|   21    |   94   |    70    |      142       |    239    |
|   26    |   65   |     8    |      169       |    310    |
|   29    |   67   |    39    |      275       |    292    |
|   32    |   86   |   111    |       27       |    204    |
|   35    |   81   |    40    |      200       |    233    |
|   37    |   81   |   199    |      281       |    167    |
|   38    |   97   |   161    |       66       |    285    |
|   39    |   53   |    46    |      186       |     19    |
|   40    |   87   |    33    |       16       |    154    |
|   41    |   64   |    57    |       82       |    181    |
|   42    |   65   |   130    |       10       |     39    |
|   45    |   76   |    33    |      116       |    268    |
|   46    |   87   |    92    |      214       |    201    |
|   48    |   90   |   177    |       78       |     18    |
+---------+--------+----------+----------------+-----------+
```

FIGURE 4.8   Detail of nodes for player B.

0.4585338519348672

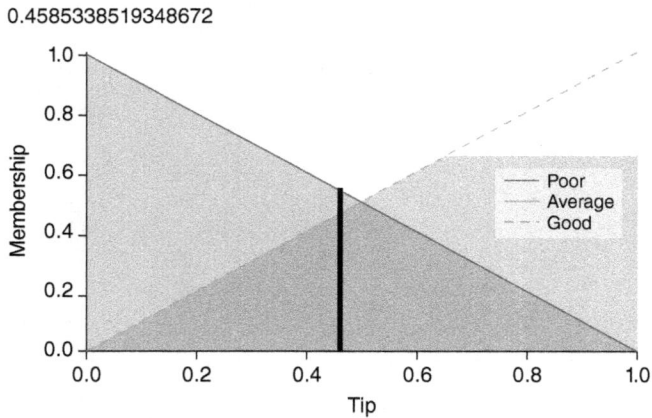

FIGURE 4.9   Membership value evaluation of fuzzy logic.

	[TEAM B] E	M	C	B
[TEAM A] E	0.43	0.5	0.33	0.49
M	0.50	0.5	0.50	0.50
C	0.50	0.5	0.33	0.49
B	0.50	0.5	0.50	0.50

FIGURE 4.10   Payoff matrix of both players.

4.6   CONCLUSIONS

The combination of fuzzy logic and game theory is used in this paper to formulate the actual work. The work is based on WSN for managing a noncooperative way to handle different situations. The complete work deals with mathematics by the fusion of membership functions of fuzzy logic. The combination of both mathematical models is used to solve fluctuations in network parameters efficiently into the payoff matrix. The proposed method does not contain any saddle point due to the nonlinear behavior of the network topology and change situations of the network parameters. Finally, the proposed method efficiently derives the optimal solutions for selecting conflicting strategies of both players in terms of probability values. These probability values help both players achieve their goals as winner and loser.

4.7   REFERENCES

[1] Arkin, E. M., Efrat, A., Mitchell, J. S., Polishchuk, V., Ramasubramanian, S., Sankararaman, S., & Taheri, J. (2014). Data transmission and base-station placement for optimizing the lifetime of wireless sensor networks. *Ad Hoc Networks*, 12, 201–218.

[2] Moh'd Alia, O. (2017). Dynamic relocation of mobile base station in wireless sensor networks using a cluster-based harmony search algorithm. *Information Sciences*, 385, 76–95.

[3] Toklu, S., & Erdem, O. A. (2014). BSC-MAC: Energy efficiency in wireless sensor networks with base station control. *Computer Networks*, 59, 91–100.

[4] Keerthika, M., & Shanmugapriya, D. (2021). Wireless sensor networks: Active and passive attacks-vulnerabilities and countermeasures. *Global Transitions Proceedings*, 2(2), 362–367.

[5] Das, S. K. (2021). Smart design and its applications: Challenges and techniques. *Nature-Inspired Computing for Smart Application Design*, 1.

[6] Dey, N. (Ed.). (2021). *Applications of Flower Pollination Algorithm and Its Variants*. Springer, Singapore, ISBN: 978-981-33-6104-1.

[7] Mahalle, P. N., Shelar, P. A., Shinde, G. R., & Dey, N. (2021). Introduction to underwater wireless sensor networks. In *The Underwater World for Digital Data Transmission* (pp. 1–21). Springer, Singapore.

[8] Keerthika, M., & Shanmugapriya, D. (2021). Wireless sensor networks: Active and passive attacks-vulnerabilities and countermeasures. *Global Transitions Proceedings*, 2(2), 362–367.

[9] Singh, A. K., Pamula, R., & Srivastava, G. (2022). An adaptive energy aware DTN-based communication layer for cyber-physical systems. *Sustainable Computing: Informatics and Systems*, 100657, https://doi.org/10.1016/j.suscom.2022.100657.

[10] Das, S. K., Maheswari, V., & Sharma, A. (2021). Wireless networks: Applications, challenges, and security issues. In *Architectural Wireless Networks Solutions and Security Issues* (pp. 1–10). Springer, Singapore.

[11] Wan, J., & Chen, J. (2022). AHP based relay selection strategy for energy harvesting wireless sensor networks. *Future Generation Computer Systems*, 128, 36–44.

[12] Singh, A. K., Pamula, R., Jain, P. K., & Srivastava, G. (2021). An efficient vehicular-relay selection scheme for vehicular communication. *Soft Computing*, 1–17.

[13] Misra, Y., Krishnaveni, K., & Rajasekaran, A. S. (2022). Implementation of NLOS based FPGA for distance estimation of elderly using indoor wireless sensor networks. *Materials Today: Proceedings*, https://doi.org/10.1016/j.matpr.2022.01.087.

[14] Singh, A. K., & Pamula, R. (2021). An efficient and intelligent routing strategy for vehicular delay tolerant networks. *Wireless Networks*, 27(1), 383–400.

[15] De, D., Mukherjee, A., Das, S. K., & Dey, N. (2020). Wireless sensor network: Applications, challenges, and algorithms. In *Nature Inspired Computing for Wireless Sensor Networks* (pp. 1–18). Springer, Singapore.

[16] Wang, F., & Hu, H. (2021). Coverage hole detection method of wireless sensor network based on clustering algorithm. *Measurement*, 179, 109449, https://doi.org/10.1016/j.measurement.2021.109449.

[17] Singh, A. K., & Pamula, R. (2021). Vehicular delay tolerant network based communication using machine learning classifiers. *Architectural Wireless Networks Solutions and Security Issues*, 195, https://doi.org/10.1007/978-981-16-0386-0_11

[18] Temene, N., Sergiou, C., Georgiou, C., & Vassiliou, V. (2022). A survey on mobility in wireless sensor networks. *Ad Hoc Networks*, 125, 102726, https://doi.org/10.1016/j.adhoc.2021.102726.

[19] Singh, A. K., Bera, T., & Pamula, R. (2018, March). PRCP: Packet replication control based prophet routing strategy for delay tolerant network. In *2018 4th International Conference on Recent Advances in Information Technology (RAIT)* (pp. 1–5). IEEE, Dhanbad, India.

[20] Das, S. K., Samanta, S., Dey, N., Patel, B. S., & Hassanien, A. E. (Eds.). (2021). *Architectural Wireless Networks Solutions and Security Issues*. Springer, Singapore.

[21] Yousefpoor, M. S., Yousefpoor, E., Barati, H., Barati, A., Movaghar, A., & Hosseinzadeh, M. (2021). Secure data aggregation methods and countermeasures against various attacks in wireless sensor networks: A comprehensive review. *Journal of Network and Computer Applications*, 103118, https://doi.org/10.1016/j.jnca.2021.103118.

[22] Zhang, J., & Mao, H. (2021). Multi-factor identity authentication protocol and indoor physical exercise identity recognition in wireless sensor network. *Environmental Technology & Innovation*, 101671, https://doi.org/10.1016/j.eti.2021.101671.

[23] Huanan, Z., Suping, X., & Jiannan, W. (2021). Security and application of wireless sensor network. *Procedia Computer Science*, 183, 486–492.

[24] Das, S. K., Dao, T. P., & Perumal, T. (Eds.). (2021). *Nature-Inspired Computing for Smart Application Design*. Springer Nature, Singapore.

[25] Shaban, W. M., Rabie, A. H., Saleh, A. I., & Abo-Elsoud, M. A. (2021). Detecting COVID-19 patients based on fuzzy inference engine and deep neural network. *Applied Soft Computing*, 99, 106906, https://doi.org/10.1016/j.asoc.2020.106906.

[26] Zahra, S. R., Chishti, M. A., Baba, A. I., & Wu, F. (2021). Detecting COVID-19 chaos driven phishing/malicious URL attacks by a fuzzy logic and data mining based intelligence system. *Egyptian Informatics Journal*, https://doi.org/10.1016/j.eij.2021.12.003.

[27] Das, S. K., Samanta, S., Dey, N., & Kumar, R. (Eds.). (2020). *Design Frameworks for Wireless Networks*. Springer, Singapore.

[28] Ahmed, H. I., Nasr, A. A., Abdel-Mageid, S. M., & Aslan, H. K. (2021). DADEM: Distributed attack detection model based on big data analytics for the enhancement of the security of internet of things (IoT). *International Journal of Ambient Computing and Intelligence (IJACI)*, 12(1), 114–139.

[29] Sholla, S., Mir, R. N., & Chishti, M. A. (2021). A fuzzy logic-based method for incorporating ethics in the internet of things. *International Journal of Ambient Computing and Intelligence (IJACI)*, 12(3), 98–122.

[30] Balusa, B. C., & Gorai, A. K. (2021). Development of fuzzy pattern recognition model for underground metal mining method selection. *International Journal of Ambient Computing and Intelligence (IJACI)*, 12(4), 64–78.

5 Cluster-Based Routing Protocol for WSN Using Fusion of Swarm Intelligence and Neural Network

Jeevan Kumar, Rajesh Kumar Tiwari, Tapan Kumar Dey, and Amit Kumar Singh

CONTENTS

5.1 Introduction 75
    5.1.1 Motivation 76
    5.1.2 Contributions 76
5.2 Literature Review 76
5.3 Proposed Method 81
5.4 Simulation and Result 86
5.5 Conclusion 89
5.6 References 89

5.1 INTRODUCTION

The wireless sensor network, or WSN, is one of the most efficient and dynamic networks for the purpose of data acquisition. It helps model several information based on sensing information and parameters. It helps model several applications that help manage data acquisition systems in the working areas [1–3]. The network consists of several components, such as base station, controller, machine monitoring system, vehicle monitoring system, medical monitoring system, large storage and management system, etc. It helps connect several information and its modeling for online monitoring systems, such as printer, server, transmitter, notebook, cellular phone, and several personal computers with sensor nodes enabling system. There are several applications of WSN, such as in forest analysis, in factories and their information management systems, in planes' information management and its related troubleshooting, in hospitals, in outdoor event systems, in military applications, in homes, in transport management, etc. It also helps in habitat monitoring,

DOI: 10.1201/b23138-7

home network management, health applications, environment observation and its application management, precision agriculture management, military application systems, pollution control and its management, home network management systems, etc. [4–6]. Although there are several applications of WSN and its related information management based on real-life application, it has also some constraints based on different network parameters. Each of the parameter affects the network lifetime based on information. The motivation and contribution of the paper follow.

5.1.1 MOTIVATION

Wireless sensor networks have evolved as a significant research subject in recent years as a result of technical advancements. Many protocols have been created to improve WSN properties, such as routing, clustering, energy, throughput, and delay, among others. In order to optimize the various parameters, we mix several sophisticated procedures.

5.1.2 CONTRIBUTIONS

a) We present a routing system based on particle swarm intelligence, ant colony optimization, and neural network properties.
b) The OMNET++ tool was used to simulate the network and analyze the results based on several characteristics, such as throughput, latency, and packet delivery ratio.
c) We compared papoose protocol to current protocols, such as PSO and ACO.

The rest of the paper is based on the following sections. The next section is based on existing works for the purpose of nature-inspired optimization and intelligent techniques. The next section deals with the proposed method for the purpose of formulation of the main goal. The next section is based on simulation analysis and the resulting discussion. The next section concludes the paper based on the work.

5.2 LITERATURE REVIEW

This literature section of the paper is used to deal with several variations of works. These works are based on the analysis and design of several types of optimizations. These optimizations are based on nature-inspired techniques for dealing with several types of information. Several works are based on several areas, including wireless network as well as wireless sensor network, or WSN. Some works are also based on COVID-19, because this is one of the current issues of most countries. Each of the work deals its contribution to paving the gap of some issues based on nature-inspired optimization. Shaban et al. [7] designed a patient analysis system that is used for inference system and design. It is based on several information that help model several neural network systems and modeling. The work is based on a fuzzy inference modeling system that helps model some issues of COVID-19. The issue is based on deep learning and fuzzy system modeling. It helps in modeling several network information based on strategy management. It helps validate several cross analyses that help in accuracy modeling. It helps model several detections for coronavirus analysis to help in modeling based on a prevention system. Zahra et al. [8] designed a method for a data-driven system that is based on intelligence system maintenance and information system. The work is based on a URL system information and modeling that helps analyze and design several malicious and phishing information systems. The work is completely based on fuzzy logic modeling. It helps deal with several uncertainty management for handling some pandemic analyses. It helps model several control and unprecedented information systems. The work helps model several cybercriminal information analyses to deal with several ransomware information systems for impact analysis. N. Dey [9] proposed a method for a nature-inspired optimization system that helps model several variant information. The work is based on several applications and modeling that help model several information systems. There are several works and applications combined in this application system that help model nonlinear optimizations efficiently and effectively. The contribution of this book helps in modeling several optimizations with the context of real-life application. It helps model several sources of information based on a multiapplication system based on power management. S. K. Das [10] proposed a method for the purpose of an application design system and its management. It helps in several application management that helps deal with several challenges and issues management. It helps model several information systems based on smart application systems. It helps model several security management systems based on emergency management and application systems. It helps give new guidelines that help model several issues in terms of solution. This solution helps model several services based on real-life application management. It helps in several monitoring and application management systems based on emergency and security modeling systems related to services. Singh et al. [11] proposed an adaptive method for the purpose of creating an energy-aware system that is used for communication purposes. This method is used for physical communication systems based on a delay-tolerant system. It helps model applications based on cyber-physical systems that are used for wireless sensor networks. It helps model applications based on delay-tolerant and prediction systems. The application is based on mobile applications for operating several operations. The work is simulated based on a network environment for predicting network lifetime. Alhasan and Hasaneen [12] designed a method based on a digital image system relating to an artificial intelligence technique for the purpose of COVID-19. The method also used the technique of natural language process for the purpose of analyzing data. It is based on public health analysis systems that use a computed tomography system. It is based on contact tracing, and its analysis system is based on an enabling system. It helps process mobile CT images for analysis of different information. The work is based on stream innovation and analysis system that helps in COVID-19 analysis and prediction systems. The work is the fusion of artificial intelligence and machine learning system for enhancing the system with model analysis. Lan et al. [13] designed a method for patient analysis and its management for the purpose of COVID-19. The method is based on an artificial intelligence system for the purpose of analyzing the coronavirus. The method is based on a public health analysis system for the purpose of controlling and managing different forecasting information. This management also uses several types of vaccines and drugs for the purpose of management. It helps reduce several challenges and accelerate different types of achievements. Tarik et al. [14] designed a method for the purpose of student performance prediction and analysis. This prediction is based on COVID-19 purposes that help manage several issues during lockdown and classes from home. The method is based on artificial intelligence, and analysis is done for students of Moroccan descent. This is one of the regions under Guelmim Oued Noun. De et al. [15] proposed an illustration for the purpose of challenges and application management services. This service is based on the application of a wireless sensor network system and its variations. It helps model several challenges and application management. It helps deal with several algorithms of wireless network based on variations and its analysis. It helps guide several working principles and information systems based on service management. It helps deal with algorithm analysis that helps manage several applications of the system. Singh et al. [16] designed an efficient modeling method for the purpose of helping in vehicular communication. The work is based on communication purposes and modeling. It helps in several application purposes that help model several vehicular relay management systems for an efficient network communication. The work is based on a utilization purpose that helps model several potential system management systems. It helps model several relay applications and network modeling. It helps predict several vehicular communications based on selfish note prediction. It helps in strategy management for tracking network performance. Haleem et al. [17] designed for cardiology analysis and prediction for COVID-19. The work is based on an artificial intelligence technique. In this work, several information is access based on the pandemic period from health care. It is based on cardiology treatment and its analysis relating to the COVID-19 disease. It helps predict and analyze several prevention techniques and interaction of information. It also helps analyze several information based on different innovation systems. It is based on a critical heart surgery system. The method uses several techniques of artificial intelligence, such as support vector machine and artificial neural network. Zhang et al. [18] designed a validation system for managing different factors of prognosis for handling different patients during the COVID-19 pandemic. The study is based on retrospective analysis based on an artificial intelligence technique. The analysis technique of this method is based on the coronavirus disease. This virus is based on a novel virus system as related to global pandemic analysis. Several techniques are used in this model for the purpose of artificial intelligence, such as artificial neural network and Lasso operator, which is based on neural network. Born et al. [19] designed an analysis model based on medical image processing for the purpose of COVID-19 analysis. This work is based on a medical image survey report for the purpose of clinical image analysis. It is based on several relevance system of images for analysis of vast recommendation. The work is also based on a systematic review of employee details and a performance system, along with patient satisfaction. It is based on several deployment and practices of clinical diagnosis. Das et al. [20] designed a book for the purpose of creating an architectural solution system based on wireless network. This service is not only based on network but also on several systems and information based on the architecture of the network. It helps in modifications based on the architecture of the system. It helps model several issues of the network. Several issues are used and deal with the system. Some of the issues mentioned are energy efficiency systems, network lifetime systems, resource management, data aggregation systems, etc. It helps in modeling several solutions and the security management of wireless networks. Singh and Pamula [21] designed a method for the purpose of creating an intelligent communication systems and route modeling. The work is based on a vehicular communication system based on strategy management. It helps, based on strategy behavior modeling and analysis, enhance network lifetime. The application is deployed for several purposes based on protocol management. It helps track analysis in novel behavior based on delay-tolerant management. It helps utilize and analyze the system based on vehicular communication. It helps outperform the results based on several variations. Jiao et al. [22] designed a patient analysis system based on prognostication for the purpose of x-ray analysis of the chest based on different clinical data and information. It is based on a retrospective analysis of different studies and management. The work is based on an artificial intelligence system and its different innovation techniques. It helps in managing and developing several clinical data and information based on different progression and its information system. The model is trained based on deep neural network analysis for the purpose of feature extraction and risk modeling system. Karaman et al. [23] designed a method for the purpose of social distance analysis for the prediction of camera analysis to analyze whether social distancing is maintained properly or not. The work is based on an artificial intelligence technique based on network analysis and predictions. The main purpose of this camera is to act as an intrusion detection system for the purpose of different types of analysis. Social distancing is maintained by convolution neural network and its analysis parameters. It uses the Raspberry Pi software for the camera and helps in different types of prediction systems. It helps in integrated system analysis for the purpose of ascertaining distance violations. Madani et al. [24] proposed a method for the purpose of fake news detection. It is based on Moroccan tweet analysis. In this model, several data are traveled and analyzed based on world information system. The work is based on several types of experimental analysis systems for detecting several types of fake news. Fake news create several types of uncertainties and ambiguities for handling several issues. It creates several types of lack of information of the model for the purpose of system analysis. Das et al. [25] proposed a book that helps model several applications based on industrial applications and formulation. It helps model several information systems based on machine learning systems and modeling. The content of this book deals with several information based on decision-making systems and prediction analysis. It helps deal with some applications based on natural language processing, machine learning, computer vision, image processing, etc. Each application and system is based on several information and modeling systems that deal with and analyze information modeling. Singh and Pamula [26] designed a delay management system for the purpose of creating a communication-based system. The work is based on machine learning–based applications that help in several purposes. The work is based on several classifier models of analysis based on solutions for vehicular networks. It helps deal with several network metrics based on certain parameters of the network. The work is based on strategy management for handling the inadequate design of a system. It helps outperform the results based on the consideration of system modeling. Maille et al. [27] designed a model based on a smartwatch electrocardiogram for the purpose of accessing cardiac rhythm safety analysis. It is based on several types of drug therapy analyses of COVID-19. It helps in managing several issues of the COVID-19 pandemic. The work is based on an artificial intelligence technique for the purpose of QT-logs information and study systems. It helps in managing several types of issues based on home and other monitoring systems. It is based on an ECG analysis system based on different variations. Yaşar et al. [28] designed a method of data and information prediction and severity analysis based on the results analysis of different protein systems. Severity is based on COVID-19 for different profile management systems using an artificial intelligence technique. The resulting analysis is based on the gradient-boosted tree, or GBT, with combination of random forest analysis. It is based on several predictive modeling and analysis systems for different effective management techniques. Mohammad-H. Tayarani N. [29] designed a method for the purpose of battling against several information and parameters of the COVID-19. The work is based on a literature review system based on the application of artificial intelligence. This review deals with several information, such as different identification systems, analysis, patient analysis and monitoring, different types of test analysis, symptoms analysis and monitoring, and several types of image analysis and testing. De et al. [30] designed a book for the purpose of wireless sensor network. It helps model several applications based on services and management. It helps deal with several information systems and in the management of key areas. The work is based on nature-inspired applications and computing that help model several issues. It helps implement several applications and computation information systems. Information from this book is distributed in the form of a bio- and nature-inspired system. It helps model and design several applications for the purpose of producing single-objective and multi-objective optimization systems. Singh et al. [31] designed a method for the purpose of strategy management based on delay analysis. It helps maintain several network management systems based on some replication analysis. It helps deal with and manage several packet control management systems. It helps in managing several data delivery and resource management systems for the purpose of communication systems handling. The work is based on routing analysis that helps deal with and manage intermediate node analysis systems. The work deals with several assists for the purpose of replication management. This replication is based on control systems and parameter-dealing systems. Li et al. [32] designed a user study system and its modeling for lung disease analysis based on different levels of severity based on chest radiograph analysis and its management. The method is based on a multiradiologist system based on an artificial intelligence technique. It is based on retrospective analysis for the purpose of creating a multiradiologist system that helps evaluate several factors of diseases. It helps analyze several interrater agreements based on its death analysis and management. It also used CXR interpretation and analysis for handling diseases efficiently. Yadav et al. [33] designed a method of biomarker-based analysis based on immunosensors from medical data analysis. It helps in managing several information based on internet of medical things, particularly those related to management with artificial intelligence. It is based on COVID-19 disease analysis to help with different types of diagnosis and management systems. The combination of artificial intelligence and internet of things related to medical information helps in managing several medical information efficiently and effectively. Kumar et al. [34] designed an illustration based on COVID-19 vaccine distribution and its analysis based on different factors and information management systems. The work of this paper is based on the fusion of two techniques, the internet of things and artificial intelligence. It helps model an application based on security, learning, and analysis [35–37]. The work is based on artificial intelligence, with real-time analysis of the health-care system and its methodologies. It incorporates several information, such as access to questions and a logistic system based on real-time application.

5.3 PROPOSED METHOD

The technique mentioned here uses the Hopfield network concept to pick the cluster head, which is followed by the nodes, which finally make up the clusters. After the clusters are found, the PSO route optimization approach is used to locate the shortest path based on optimal network lifetime. The proposed method is divided into sections for designing a cluster-based route.

(a) **Hopfield Network.** The equation below shows the activation function of each neuron, which is calculated as threshold function, as shown in equation 5.1, and a very simple Hopfield network, as shown in Figure 5.1.

$$\forall u \epsilon U : f_{act}^{(u)}\left(net_u, \theta_u\right) = \begin{cases} 1, if\ net_u \geq \theta, \\ -1, otherwise. \end{cases} \tag{5.1}$$

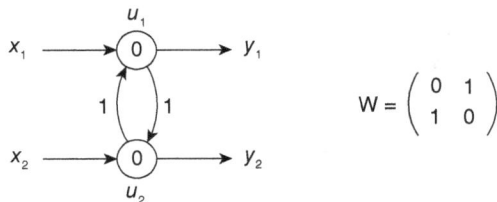

FIGURE 5.1 Hopfield network.

The Hopfield network behavior depends on two formations of neurons, like parallel and sequential. In the first case, computation is based on oscillation, and in the second case, computation is based on convergence.

(i) Parallel update of neurons, as shown in Figure 5.2.

	u_1	u_2
input phase	−1	1
work phase	1	−1
	−1	1
	1	−1
	−1	1
	1	−1
	−1	1

FIGURE 5.2   Parallel update of neuron activation.

- With the context of computations on the oscillate system, there is no stable state reached. So computation is terminated based on dependent output.

(ii) Neurons updated sequentially, as is shown in Figure 5.3.

	u_1	u_2
input phase	−1	1
work phase	1	1
	1	1
	1	1
	1	1

	u_1	u_2
input phase	−1	1
work phase	−1	−1
	−1	−1
	−1	−1
	−1	−1

FIGURE 5.3   Sequential update of neuron activation.

- A stable state is reached based on the order of the update. The update order is maintained by reaching the state of the system.

Assume that all nodes are normal. When a node is designated as a cluster head, the cluster head can be chosen based on the existing CHs percentage. We also examine how many times the same sensor node is marked as CH, and then we examine the sensor node's power life. Prior to cluster head selection, the one hopfield concept of neural networks is used to examine the eligible set of sensor nodes to become CH. Hopfield networks are used for pattern recognition. Weights are assigned to each node and must have a value of +1 or −1.

$$T = [\ +1 \ +1 \ −1 \ +1; \ldots\ldots$$
$$−1 \ +1 \ +1 \ −1; \ldots\ldots$$
$$+1 \ +1 \ +1 \ +1; \ldots\ldots$$
$$−1 \ −1 \ +1 \ +1 \ \ ]$$

The first parameter indicates the amount of energy consumed, the second parameter indicates the distance, the third parameter indicates the neighbors, and the fourth parameter indicates the ratio of the current energy to the remaining energy. In the best-case scenario, the values are [-1–1 +1 +1], while in the worst-case scenario, they are [+1 +1–1 -1]. For CH selection, the sensor nodes with the best and average set of values are chosen.

(b) **Cluster Formation.** Finally, after equating the eligible set of nodes, the average power source for all nodes is computed, which is referred to as the threshold value (d0). Only nodes with an energy level equal to or greater than the threshold value are allowed to participate in CH selection. Nodes that fail to meet the required condition for participation in the CH selection process must wait for 1/pr rounds. Eligible nodes generate a random number, ranging from 0 to 1. To become a CH, this number must be less than the threshold Th(nd). If this number is less than the threshold Th(nd), the node becomes the one that meets the condition, and the label *CH* is assigned to it. Equation 5.2 gives the threshold value.

$$Th(nd) = \begin{cases} \dfrac{pr}{1 - pr * \left(roumod\dfrac{1}{pr}\right)}, nd \in Gr \\ 0, otherwise \end{cases} \tag{5.2}$$

Where *pr* represents the percentage of CH, *rou* represents the current round, and *Gr* represents the set of nodes that are not CH. The remaining nodes that do not become CH and RN will function as normal nodes (NN).

Once all CHs and RNs have been chosen, these CHs and RNs broadcast a message to the NNs. At this point, the receivers of the NNs must be "on" in order to hear the message broadcasted by the CHs and RNs. The NNs choose their cluster and look for RNs that are close to it. This decision is made based on the distance between CHs and RNs. The signal strength is used to calculate the distance; the greater the signal strength, the shorter the distance. NN will attach to the cluster head with the shortest distance.

(c) **Swarm Intelligence Algorithms.** After the cluster schedule has been created and established, the following step is to discover the shortest data transmission paths in order to reduce energy consumption. As a result, the PSO optimization technique is used to determine the shortest path from node to sink [35]. The results are refined using particle swarm optimization (PSO).

Data transmission begins after the quickest paths have been calculated. Assume that all nodes in each round have some data and that they all generate data at the same rate. Normal nodes submit their data to CH throughout the period allotted to them. In order to save electricity, NN must keep their radio turned on for the entire time period permitted. During this time, all CH and RN receivers must be turned on. After collecting data from all nodes, CH begins aggregating the information and sending it to either the mobile sink or the nearest RN.

(d) **Proposed Algorithm.** Algorithm 1 indicates the details of the proposed method, and the cluster head selection mechanism is shown in pseudocode 1. The proposed architecture is shown in Figure 5.4, and the pseudocode of neural network and the PSO algorithm for route selection are shown in pseudocode 2 and pseudocode 3.

Algorithm 1: The proposed method.

Step 1. Initialized all nodes.
Step 2. After initializing, nodes are deployed in WSNs.
Step 3. The selection of cluster head is based on neural network technique.
Step 4. Path analysis for base station and cluster head is determined by PSO technique.
Step 5. Data transmission is performed based on the operation.
Step 6. During optimized path analysis, remaining energy is calculated for each of the nodes in the network.
Step 7. The network lifetime is determined.
Step 8. End.

Pseudocode 1: Cluster head selection method.

1. N← Number of nodes that are used in the targeted area.
2. // Initialize the sensors in the deploying area.
3. S ← {n_1, n_2, ... n_N} // Set of all active nodes in the field.
4. Let i←0.
5. For each *i* in the set S
       E_i ← Set Residual energy.
6. End for.
7. No_of_rounds ← 1000 // set no of rounds.
8. T← Set threshold energy.
9. For each i in No_of_rounds
       Cluster_head ← Algo_nuralnetwork(S, E)
           For each j in the set S
           E_j ← update Residual energy.
           End for.
       If (E_j< T) // Comparing the residual energy with threshold.
         Discard the node from the set.
       Else,
           Continue.
       End for.

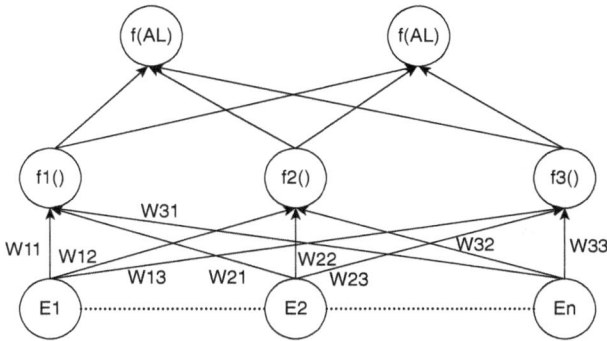

FIGURE 5.4 The proposed architecture.

Pseudocode 2: Neural network algorithm.

Algo_nurralnetwork():

1. X← [E0, E1, E2 . . . . En-1] // Input vector composes the node residual energy.
2. Y ← [Y0, Y1, Y2, ... Ym-1] //The output vector contains the probability of each node to become a cluster head for each round.
3. L← no of hidden layes.
4. W_i and b_i are the weight and bias between (i-1) and i, where $(0 < I < L)$.
5. For each hidden layer i
        For each neuron j
                $A_{ij}(X) \leftarrow b_i + \Sigma W_{ij} * h_{ij}$ The preactivation at the ith layer, and j is the number of neuron in layer i

                $h_{ij} \leftarrow f(A_{ij}(X))$ where $f(A_{ij}) \leftarrow \dfrac{1}{1+e^{A_{ij}}}$
6. End for.
7. End for
8. $A_L \leftarrow$ Preactivation at the output layer.
9. $Y \leftarrow f(A_L)$
10. // Choose the softmax function as activation function at the output layer.
11. $Y \leftarrow \dfrac{e^{ALJ}}{\sum_{j=1}^{m} e^{ALj}}$
12. CH ← Y_{max} // pick the node with highest probability and return.
13. Return (CH) // Cluster head.

Pseudocode 3: PSO algorithm for route selection.
1. Initialize the nodes in the sensor field.
2. Find out potion and velocity of the sensor node.
3. Create different path between nodes.
4. Select the shortest path.

(e) **Fuzzy Logic for Modeling Cluster.** The modeling of the network also uses fuzzy logic for controlling different types of imprecise information. During cluster selection, several types of issues are raised due to continuous changes in the network. It creates some confusion to make an efficient cluster. So fuzzy logic is used to turn the network parameters into linguistic variables. This linguistic variable is designed by a membership function, like the triangular membership function. It helps model the network as a robust connection by combining nature-inspired optimization, such as swarm intelligence and artificial neural network. Hence, it efficiently selects better clusters based on the required operation in the network.

5.4 SIMULATION AND RESULT

The suggested scheme's performance in various simulated scenarios is depicted in this section. For the simulation, the OMNET++ tool was utilized, which supports routing protocols for wireless sensor networks. OMNET++ is a discrete, event-driven network simulator that is object-oriented. Because of its simplicity and adaptability, the OMNET++ network simulator has garnered tremendous appeal among members of the research community. The network simulation allows simulation scripts, also known as simulation scenarios, to be easily created in a language, while still relying on more complex capabilities via C++ code. Several metrics are used to evaluate performance, as shown in Figures 5.5 to 8. All nodes

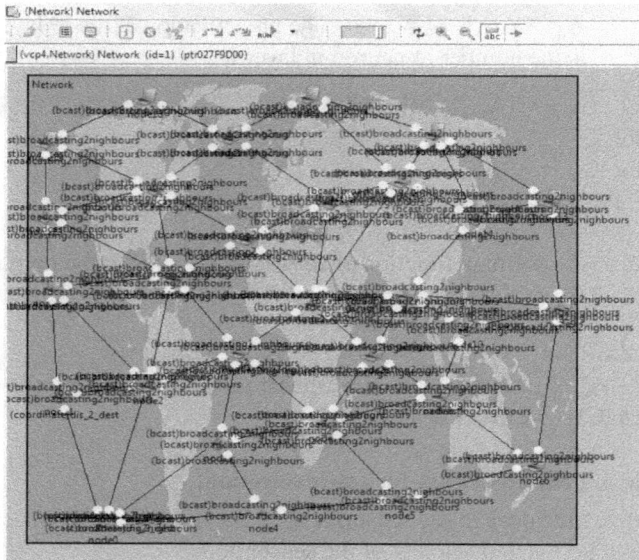

FIGURE 5.5 Gathering information of sensor network.

broadcast a message to each other and retain surrounding node information, as shown in Figure 5.5. This data is saved in a routing table. When transferring data packets, this method assists in determining which node is closest to the target node. First, define the source and destination nodes; next, from the routing table and determine the surrounding node information. Compare all nodes with their relevant parameters to determine which is closest to the destination. The data packet is transmitted to the appropriate node that is closest to the destination node. Repeat this process until all data packets have arrived at their destination.

Figure 5.6 shows the figure drawn between the throughput rate and the node number. We must now compare the average amount of data received by the network's destination nodes per unit time. Then, compare the throughput with respect to the corresponding node number of PSO, ACO, and PSO-NN. Finally, PSO-NN protocol has higher throughput rate than the existing protocols.

Figure 5.7 shows this discussion about packet delivery ratio. We find the packet delivery ratio of PSO, ACO, and PSO-NN. The packet delivery ratio in a network is the ratio of the total number of data packets received by destination nodes to the total number of data packets generated or sent by source nodes.

The average end-to-end delay of the PSO-NN, PSO, and ACO protocols is depicted in Figure 5.8. The average end-to-end delay defines the amount of time it takes for all data packets in the network to travel from source nodes to destination nodes. In this figure, delay is minimum from source to destination through PSO-NN protocol.

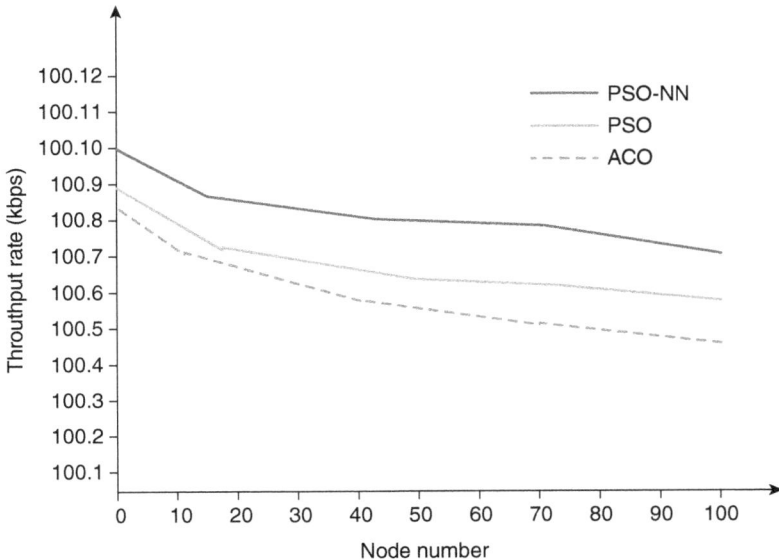

FIGURE 5.6 Throughput comparison of different existing protocols, such as PSO-NN, PSO, and ACO.

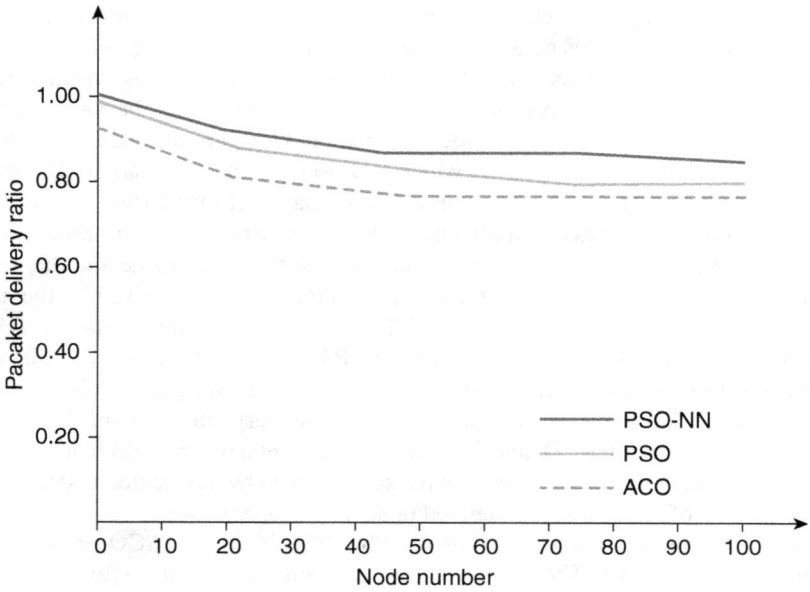

FIGURE 5.7 Packet delivery ratio comparison of different existing protocols, such as PSO-NN, PSO, and ACO.

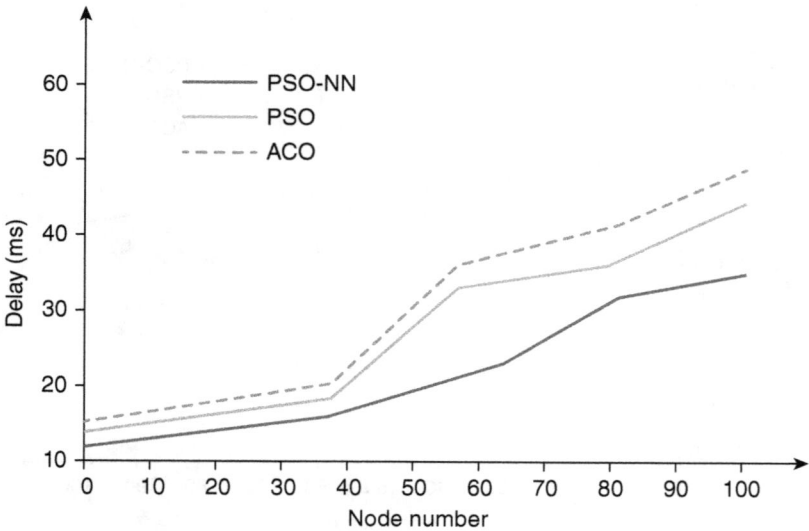

FIGURE 5.8 Delay comparison of different existing protocols, such as PSO-NN, PSO, and ACO.

5.5 CONCLUSION

Finding ways to reduce power usage is one of the most serious challenges with WSNs. For WSNs, we developed an enhanced NN-based PSO routing method. By minimizing transmission distance, the proposed work improves the CH determination and selection process, and the PSO algorithm aids in determining the shortest route and path, resulting in decreased power consumption and increased network lifetime. Simulations show that the proposed technique is beneficial in terms of energy reduction in both scenarios, even when the number of nodes or network size is increased; in both cases, the proposed algorithm works to extend the network lifetime.

5.6 REFERENCES

[1] Keerthika, M., & Shanmugapriya, D. (2021). Wireless sensor networks: Active and passive attacks-vulnerabilities and countermeasures. *Global Transitions Proceedings*, 2(2), 362–367.

[2] Forero, F., & da Fonseca, N. L. (2022). Distribution of multi-hop latency for probabilistic broadcasting protocols in grid-based Wireless Sensor Networks. *Ad Hoc Networks*, 126, 102754, https://doi.org/10.1016/j.adhoc.2021.102754.

[3] Binh, H. T. T., Hanh, N. T., Van Quan, L., & Dey, N. (2018). Improved cuckoo search and chaotic flower pollination optimization algorithm for maximizing area coverage in wireless sensor networks. *Neural Computing and Applications*, 30(7), 2305–2317.

[4] Rashid, B., & Rehmani, M. H. (2016). Applications of wireless sensor networks for urban areas: A survey. *Journal of Network and Computer Applications*, 60, 192–219.

[5] Karaboga, D., & Akay, B. (2009). A survey: Algorithms simulating bee swarm intelligence. *Artificial Intelligence Review*, 31(1–4), 61.

[6] Binh, H. T. T., & Nam, N. H. (2018). Introduction to coverage optimization in wireless sensor networks. In *Soft Computing in Wireless Sensor Networks* (pp. 115–136). Chapman and Hall/CRC, London.

[7] Shaban, W. M., Rabie, A. H., Saleh, A. I., & Abo-Elsoud, M. A. (2021). Detecting COVID-19 patients based on fuzzy inference engine and Deep Neural Network. *Applied Soft Computing*, 99, 106906, https://doi.org/10.1016/j.asoc.2020.106906.

[8] Zahra, S. R., Chishti, M. A., Baba, A. I., & Wu, F. (2021). Detecting COVID-19 chaos driven phishing/malicious URL attacks by a fuzzy logic and data mining based intelligence system. *Egyptian Informatics Journal*, https://doi.org/10.1016/j.eij.2021.12.003.

[9] Dey, N. (Ed.). (2021). *Applications of Flower Pollination Algorithm and Its Variants*. Springer, Singapore, ISBN: 978-981-33-6104-1.

[10] Das, S. K. (2021). Smart design and its applications: Challenges and techniques. *Nature-Inspired Computing for Smart Application Design*, 1.

[11] Singh, A. K., Pamula, R., & Srivastava, G. (2022). An adaptive energy aware DTN-based communication layer for cyber-physical systems. *Sustainable Computing: Informatics and Systems*, 100657, https://doi.org/10.1016/j.suscom.2022.100657.

[12] Alhasan, M., & Hasaneen, M. (2021). Digital imaging, technologies and artificial intelligence applications during COVID-19 pandemic. *Computerized Medical Imaging and Graphics*, 101933, https://doi.org/10.1016/j.compmedimag.2021.101933.

[13] Lan, L., Sun, W., Xu, D., Yu, M., Xiao, F., Hu, H., & Wang, X. (2021). Artificial intelligence-based approaches for COVID-19 patient management. *Intelligent Medicine*, https://doi.org/10.1016/j.imed.2021.05.005.

[14] Tarik, A., Aissa, H., & Yousef, F. (2021). Artificial intelligence and machine learning to predict student performance during the COVID-19. *Procedia Computer Science*, 184, 835–840.

[15] De, D., Mukherjee, A., Das, S. K., & Dey, N. (2020). Wireless sensor network: Applications, challenges, and algorithms. In *Nature Inspired Computing for Wireless Sensor Networks* (pp. 1–18). Springer, Singapore.

[16] Singh, A. K., Pamula, R., Jain, P. K., & Srivastava, G. (2021). An efficient vehicular-relay selection scheme for vehicular communication. *Soft Computing*, 1–17.

[17] Haleem, A., Javaid, M., Singh, R. P., & Suman, R. (2021). Applications of artificial intelligence (AI) for cardiology during COVID-19 pandemic. *Sustainable Operations and Computers*, 2, 71–78.

[18] Zhang, S., Huang, S., Liu, J., Dong, X., Meng, M., Chen, L., & Chen, D. (2021). Identification and validation of prognostic factors in patients with COVID-19: A retrospective study based on artificial intelligence algorithms. *Journal of Intensive Medicine*, https://doi.org/10.1016/j.jointm.2021.04.001.

[19] Born, J., Beymer, D., Rajan, D., Coy, A., Mukherjee, V. V., Manica, M., & Rosen-Zvi, M. (2021). On the role of artificial intelligence in medical imaging of COVID-19. *Patterns*, https://doi.org/10.1016/j.patter.2021.100269.

[20] Das, S. K., Samanta, S., Dey, N., Patel, B. S., & Hassanien, A. E. (Eds.). (2021). *Architectural Wireless Networks Solutions and Security Issues*. Springer, Singapore.

[21] Singh, A. K., & Pamula, R. (2021). An efficient and intelligent routing strategy for vehicular delay tolerant networks. *Wireless Networks*, 27(1), 383–400.

[22] Jiao, Z., Choi, J. W., Halsey, K., Tran, T. M. L., Hsieh, B., Wang, D., & Bai, H. X. (2021). Prognostication of patients with COVID-19 using artificial intelligence based on chest x-rays and clinical data: A retrospective study. *The Lancet Digital Health*, 3(5), e286–e294.

[23] Karaman, O., Alhudhaif, A., & Polat, K. (2021). Development of smart camera systems based on artificial intelligence network for social distance detection to fight against COVID-19. *Applied Soft Computing*, 110, 107610, https://doi.org/10.1016/j.asoc.2021.107610.

[24] Madani, Y., Erritali, M., & Bouikhalene, B. (2021). Using artificial intelligence techniques for detecting COVID-19 epidemic fake news in Moroccan tweets. *Results in Physics*, 25, 104266, https://doi.org/10.1016/j.rinp.2021.104266.

[25] Das, S. K., Das, S. P., Dey, N., & Hassanien, A. E. (Eds.). (2021). *Machine Learning Algorithms for Industrial Applications*. Springer, Switzerland.

[26] Singh, A. K., & Pamula, R. (2021). Vehicular delay tolerant network based communication using machine learning classifiers. *Architectural Wireless Networks Solutions and Security Issues*, 195, https://doi.org/10.1007/978-981-16-0386-0_11

[27] Maille, B., Wilkin, M., Million, M., Rességuier, N., Franceschi, F., Koutbi-Franceschi, L., & Fiorina, L. (2021). Smartwatch electrocardiogram and artificial intelligence for assessing cardiac-rhythm safety of drug therapy in the COVID-19 pandemic. The QT-logs study. *International Journal of Cardiology*, 331, 333–339.

[28] Yaşar, Ş., Çolak, C., & Yoloğlu, S. (2021). Artificial intelligence-based prediction of COVID-19 severity on the results of protein profiling. *Computer Methods and Programs in Biomedicine*, 202, 105996, https://doi.org/10.1016/j.cmpb.2021.105996.

[29] Tayarani-N, M. H. (2020). Applications of artificial intelligence in battling against COVID-19: A literature review. *Chaos, Solitons & Fractals*, 110338, https://doi.org/10.1016/j.chaos.2020.110338.

[30] De, D., Mukherjee, A., Das, S. K., & Dey, N. (Eds.). (2020). *Nature Inspired Computing for Wireless Sensor Networks*. Springer, Singapore.

[31] Singh, A. K., Bera, T., & Pamula, R. (2018, March). PRCP: Packet replication control based prophet routing strategy for delay tolerant network. In *2018 4th International Conference on Recent Advances in Information Technology (RAIT)* (pp. 1–5). IEEE, Dhanbad, India.

[32] Li, M. D., Little, B. P., Alkasab, T. K., Mendoza, D. P., Succi, M. D., Shepard, J. A. O., & Kalpathy-Cramer, J. (2021). Multi-radiologist user study for artificial intelligence-guided grading of COVID-19 lung disease severity on chest radiographs. *Academic Radiology*, 28(4), 572–576.

[33] Yadav, A. K., Verma, D., Kumar, A., Kumar, P., & Solanki, P. R. (2021). The perspectives of biomarkers based electrochemical immunosensors, artificial intelligence and the internet of medical things towards COVID-19 diagnosis and management. *Materials Today Chemistry*, 100443, https://doi.org/10.1016/j.mtchem.2021.100443.

[34] Kumar, R., & Veer, K. (2021). How artificial intelligence and internet of things can aid in the distribution of COVID-19 vaccines. *Diabetes & Metabolic Syndrome*, https://doi.org/10.1016/j.dsx.2021.04.021.

[35] Ahmed, H. I., Nasr, A. A., Abdel-Mageid, S. M., & Aslan, H. K. (2021). DADEM: Distributed attack detection model based on big data analytics for the enhancement of the security of internet of things (IoT). *International Journal of Ambient Computing and Intelligence (IJACI)*, 12(1), 114–139.

[36] Sholla, S., Mir, R. N., & Chishti, M. A. (2021). A fuzzy logic-based method for incorporating ethics in the internet of things. *International Journal of Ambient Computing and Intelligence (IJACI)*, 12(3), 98–122.

[37] Balusa, B. C., & Gorai, A. K. (2021). Development of fuzzy pattern recognition model for underground metal mining method selection. *International Journal of Ambient Computing and Intelligence (IJACI)*, 12(4), 64–78.

6 Nonlinear Fuzzy Optimization Technique for WSN Based on Quadratic Programming

Manoj Kumar Mandal, Arun Prasad Burnwal,
B. K. Mahatha, and Abhishek Kumar

CONTENTS

6.1 Introduction 93
6.2 Related Works 94
6.3 Proposed Method 98
6.4 Simulation and Analysis 103
6.5 Conclusion 111
6.6 References 111

6.1 INTRODUCTION

A wireless sensor network, or WSN, is one type of intelligent network which is used for network optimization based on several parameters. It is used in several applications based on a base station, or BS, service management. It contains several types of nodes, such as the normal node, sink node, source node, destination node, etc., for managing different types of services [1–3]. The purpose of the sensor node is to sense environmental information and send it to the BS based on user requirement. The BS collects this information and processes it for future references. The applications of WSN spreads all over in human life, such as in entertainment, military applications, communications, disaster management, cloud computing, education system, hospital management, hotel management, etc. The network consists of several benefits and applications, but it has also some limitations, like its limited battery constraints [4]. It is used in several services, but due to this limitation, it breaks the services and fails the operation. Some of the issues mentioned are data loss, route or communication breaks, unpredictability, route fails, mission or operation fails, etc.

Therefore, in this paper, fuzzy logic is used with nonlinear optimization to deal with several types of operations efficiently. Fuzzy logic works with fuzzy linguistic variable for dealing with several types of service. This linguistic variable is

DOI: 10.1201/b23138-8

distributed by membership function with input parameters to control several types of uncertainties. This input parameter, along with fuzzy linguistic variable, helps model objective function. This objective function, along with some constraints, helps model the application based on services. Finally, it helps to achieve the main goal of the services.

The remainder of the paper is divided into several sections. The next section is used to deal with several works for the purpose of optimization. The next section is used to define several information based on the proposed method. The next section analyzes simulation parameters and its performance evaluation system. The last section is used to conclude the paper based on the goal.

6.2 RELATED WORKS

In the last decade, several works have been proposed with the same context. Most of the works are based on network formulation and optimization that are discussed in this section as Das et al. [5] designed an application management illustration for the purpose of wireless network system and services. It helps model several applications based on different security and challenge system management. It helps in issue management based on several parts of the communication system. It helps adopt several solutions based on some higher analysis and management. The work is based on complexity management, which helps model several solutions within the context of management based on different variations of the wireless network. N. Dey et al. [6] designed a method for the purpose of big data analysis and modeling. It helps in modeling several next-generation information systems. It helps model several intelligence applications based on different types of services. The works of this book are based on the fusion of the internet of things, cloud computing, and several intelligence systems. It helps model new techniques and services for modeling several services based on new and smart applications. It helps model several big data analytics that help model some internet of things information systems. Singh and Pamula [7] designed a method for the purpose of intelligent communication and route modeling. The work is based on vehicular communication systems based on strategy management. It helps, based on strategy behavior modeling and analysis, enhance network lifetime. The application is employed in several purposes based on protocol management. It helps track analysis in novel behavior based on delay-tolerant management. It helps utilize and analyze the system based on vehicular communication. It helps outperform the result based on several variations. N. Dey [8] proposed several methods and operations based on firefly algorithms that help design several types of services. It helps model several embedding and medical analysis systems. The work is based on applications and management that help model several imaging systems. The works of this book are based on some application management systems that help in managing several metaheuristic systems. It helps in managing several decomposition analyses based on image analysis. This analysis is based on the fusion of several embedded information systems. It helps in managing several electronic patient record information systems based on nature-inspired systems and modeling. Keerthika and Shanmugapriya [9] designed a method with the combination of passive and active attacks for the purpose of illustration. This illustration is based on some countermeasures system that helps in vulnerabilities system. It helps model the application based on environment analysis, which helps deal with a protection system based on commercial analysis. The communication of the system is based on the deployment of some challenges along with some issues. It helps in defensive analysis based on vulnerability analysis of some factors of information. Das et al. [10] designed a book for the purpose of smart application design. The work of this book is based on smart application with the fusion of smart computing. It helps in modeling several applications based on some services and on management. It helps model several nature-inspired applications based on computing application services. It helps model and give new sight in the subarea of network modeling, data analysis and prediction, network lifetime management, resource and energy management, etc. It helps model several information systems for the management of dynamic application and planning and services. Prasad and Shivashankar [11] designed an enhanced protocol system for ad hoc network. In this system, the network is based on mobile ad hoc network that helps design a routing system. It helps in managing several challenges and routing information system for management and its analysis. It helps in managing several source nodes and its information system based on target node analysis. The work is based on policy management that helps communicate the system efficiently. It helps in managing systems based on an autonomous system that helps in managing several network lifetimes. Wan and Chen [12] designed a strategy for energy analysis and mechanism for harvesting analysis. The work is based on the WSN purpose of modeling. It helps model several cooperative analyses for node analysis. It defines some probability based on relay node detection. The main purpose of this analysis is to solve network performance based on certain factors. It helps model the application and save the actual energy. It uses mathematical modeling for analyzing data and parameters. It helps enhance the energy based on solar energy system and its cooperation. Singh and Sharma [13] designed a process for a routing system that helps model the mobile ad hoc network. It is based on the optimization process that helps make the operation resilient. It also helps monitor the environment based on vehicular ad hoc network. The work helps model several device-to-device communication systems based on the requirement. The outcome of this application is deployed in several areas based on service management. The work is optimized based on nature-inspired optimization for handling several eliminations. S. K. Das [14] proposed a method for the purpose of creating an application design system and its management. It helps in several application management systems that help deal with several challenges and issue management. It helps model several information systems based on a smart application system. It helps model several security management systems based on emergency management and application systems. It helps give new guidelines that help model several issues in terms of solution. This solution helps model several services based on real-life application management. It helps in several monitoring and application management systems based on an emergency and security modeling system based on services. Misra et al. [15] designed an implementation method based on the fusion of FPGA and NLOS. The work is based on distance analysis and its estimation system. It helps model several applications that help the elderly mode. The work is designed for the purpose of creating an indoor system that helps in WSN. It helps in location analysis based on the ZigBee network. The work uses a programmable gate array system and its modeling. This modeling uses artificial neural network to estimate different errors and improve network lifetime. It uses a hybridization method for modeling several complexities based on suitable analysis. Anand et al. [16] designed a framework system for managing several applications based on multicast service. It helps in managing several protocol systems based on service management. The work is based on dependability analysis and productivity management, which help in single transmission. The work helps model several retreating systems to manage transparency. It helps deal with several confident and legitimate analysis to manage node along with network information. It helps in managing several intrusions and in increasing network lifetime. Dey et al. [17] designed a method for wireless sensor network application management that helps model several information based on a health-care information system. The work of this information system helps develop residential information systems based on ECG application. It helps model several health-care information and monitoring systems. The contribution of this application is based on some automation and industry management. It helps model several electrical activities that help measure and monitor information systems. Wang and Hu [18] designed a hole-detection method for handling several issues based on WSN. The network is based on a clustering method and algorithm that uses some gap coverage analysis. It helps analyze multihop management systems for rational deployment. It helps distance parameters system and vulnerability detection that help in coverage and its parameter modeling. It overcomes the limitation of several determination systems for edge node modeling. It helps determine random walk connection and its management. Tahir et al. [19] designed a clustering system to manage several communication systems based on peer-to-peer network and its management. The work helps deal with several overlay management systems based on applications. The clustering system of the network is based on multiple dimensional analysis. It helps in managing several linkages and its analysis, which help lookup management. It helps decrease several complexities, such as overhead, computation system, error, etc. Finally, it helps model several environment analyses based on path management. Temene et al. [20] illustrated a survey based on mobility analysis and prediction for WSN. The work is based on IoT and WSN both for a detailed illustration. It helps model several mobile nodes. There are several mobile nodes that play different roles, such as the sink node, mobile node, source node, etc. The combination of all nodes helps model several congestions and their related mitigations. It helps in the predecessor analysis of IoT, which helps in several directions. The work helps in modeling several evaluations based on different algorithms. De et al. [21] proposed an illustration for the purpose of challenges and application management services. This service is based on the application of a wireless sensor network system and its variations. It helps model several challenges and application management systems. It helps deal with several algorithms of the wireless network based on variation and its analysis. It helps guide several working principles and information systems based on service management. It helps deal with algorithm analysis, which helps in managing several applications of the system. Sharma and Kim [22] designed a method for multipath management that helps model several information. It helps model several applications based on routing information that helps model a network. This network is based on the application management of mobile ad hoc networks. The work uses a bioinspired technique that helps in managing some applications based on services. It helps model several constraints, such as low memory management, bandwidth, battery life, etc. The combination of all information helps model several applications along with services to adjust the model of the network. Yousefpoor et al. [23] designed a secure method for WSN as review paper that helps model several issues in the network. The work is based on a data aggregation method that helps reduce the attack in the system. It helps in countermeasures for several issues within the context of attack measurement. This review is also based on the industrial internet of things system and its modeling. It helps in managing several issues with the context of solution measurement. It helps save energy and increase security of the system based on an authentication system. Das et al. [24] designed a book for the purpose of architectural solution system based on wireless networks. This service is not only based on network but also on several systems and information based on the architecture of the network. It helps in modifications based on the architecture of the system. It helps model several issues of the network. Several issues are used and deal with the system. Some of the issues mentioned include the energy efficiency system, network lifetime system, resource management, data aggregation system, etc. It helps model several solutions and the security management of wireless networks. Lee et al. [25] designed a technique that helps model several information-sharing systems. The work is based on a military application that uses a mobile ad hoc network that helps in managing some applications. The proposed work is termed a cooperative phase with a steering system based on relay management. It consists of relay and destination node management based on services. The network also attaches with a cognitive network model to increase the probability of routing. It helps select relay based on several source nodes to increase the performance of the model. Singh et al. [26] designed an efficient modeling method to help in vehicular communication. The work is based on communication purposes and modeling. It helps in several applications that help model several vehicular relay management systems for the purpose of creating efficient network communication. The work is based on a utilization purpose that helps model several potential system management. It helps model several relay applications and network modeling. It helps predict several vehicular communication systems based on selfish note prediction. It helps in strategy management for tracking network performance. De et al. [27] designed a book for the purpose of wireless sensor network. It helps model several applications based on services and management. It helps deal with several information systems and in the management of key areas. The work is based on a nature-inspired application and computing that helps model several issues. It helps implement several applications and computation information systems. Information from this book is distributed in the form of a bio- and nature-inspired system that helps in modeling security analysis and its learning methodology [28–30]. It helps model and design several applications for single-objective and multiobjective optimization systems.

6.3 PROPOSED METHOD

The suggested approach for WSN routing is based upon energy-efficient routing. In this study, two fundamental parameters are used: residual energy and node distance. The remaining energy is the node's leftover energy; however, at the beginning, the initial energy and remaining energy are comparable. In this WSN, the nature of the network parameters are two types, given in equations 1 and 2.

$$\text{Nature1} = \overline{Residual\_Energy} \cong \text{Network\_Lifetime} \tag{1}$$

$$\text{Nature2} = \overline{Distance} \cong \text{Network\_Lifetime} \tag{2}$$

Equation 1 indicates that if residual energy is increased, then network lifetime of the WSN also increases. It means the nature of the residual energy and the nature of the network lifetime are both the same. The parameters "residual energy" and "energy consumption" both are different. *Residual energy* indicates the remaining energy of the node, and *energy consumption* indicates consuming energy of the node. If residual energy is increased, then network lifetime is increased, but if energy consumption is increased, then network lifetime will decrease. Equation 2 indicates that if distance is increased between source and destination nodes, then network lifetime is decreased. Its means the nature of the distance and network lifetime is contradictory. Equations 3 to 5 are considered a set of sensor nodes, set of energy consumption, and set of distances.

$$\text{Set1} = \{n1, n2, n3, n4, \dots \dots \dots \dots \dots \dots \dots \dots \dots \dots \dots \dots ., nj\} \tag{3}$$

$$\text{Set2} = \{e1, e2, e3, e4, \dots \dots \dots \dots \dots \dots \dots \dots \dots \dots \dots \dots ., ek\} \tag{4}$$

$$\text{Set3} = \{d1, d2, d3, d4, \dots \dots \dots \dots \dots \dots \dots \dots \dots \dots \dots \dots ., dl\} \tag{5}$$

Where the values of j, k, and l may not be equal.

Table 6.1 shows the environment of sensor nodes randomly deployed, which consists of several node IDs, axis of the sensor nodes, which consists of X and Y coordinates, and radio range of the sensor nodes, which gives access from one sensor node to another sensor node. The environment of the WSN is dynamic— that is, it changes from time to time based on the requirement. So the parameters of the network also change frequently. In this proposal, energy is considered as

100 J and distance is considered as 500 meters. Due to variations of the network, the energy parameter varies based on the fuzzy membership functions mentioned in Table 6.2. Same for distance parameter, which varies based on fuzzy membership functions based on what is mentioned in Table 6.3.

The major goal of this study is to decrease the energy consumption and distance of the sensor nodes along any path. As a result, the goal function and its limitations are illustrated in equations 6 to 11. In these equations, g_1 and g_2 are objectives of energy consumption and distance, where the combination of both is minimized. In this proposal, the total number of linguistic variables is 3 for each parameter.

TABLE 6.1
The Environment of the Nodes

Node ID	(X, Y)	Radio Range
n_1	x_1, y_1	r_1
n_2	x_2, y_2	r_2
n_3	x_3, y_3	r_3
n_4	x_4, y_4	r_4
...	...	...
...	...	...
...	.....	...
n_{j-2}	x_{j-2}, y_{j-2}	r_{j-2}
n_{j-1}	x_{j-1}, y_{j-1}	r_{j-1}
n_j	x_j, y_j	r_j

TABLE 6.2
Membership Functions of Energy

Linguistic variable	Notation	Range
Low	E_L	(0–30)
Medium	E_M	(20–75)
High	E_H	(60–100)

TABLE 6.3
Membership Functions of Distance

Linguistic variable	Notation	Range
Short	D_S	(0–200)
Medium	D_M	(180–350)
Long	D_L	(340–500)

Linguistic variables for energy are "low," "medium," and "high," and the linguistic variables for distance are "short," "medium," and "long." Equations 6 to 8 are design objective functions for three linguistic variables of energy and distance based on 100 nodes in quadratic form. Equations 9 to 11 are design objective functions for three linguistic variables of energy and distance based on 200 nodes in the form of quadratic programming. Quadratic programming is an optimization approach, similar to metaheuristic [30], for optimizing a variety of problems depending on objective function and constraints. As it is more effective than linear programming, it is employed in the proposed method. The proposed work is validated into two variations, i.e., variation 1 and variation 2, under two rounds, where the first round contains 100 nodes and the second round contains 200 nodes. Tables 6.4 to 6.6 show details of dataset under nodes 100, and Tables 6.7 to 6.9 show details of dataset under nodes 200. In these datasets, Obj_i are different objectives that varies $i = 1$ to 6 for two rounds, 100 and 200, for two variations each under three linguistic variations for energy consumption and distance.

TABLE 6.4
Dataset for Round 1 Under 100 Sensor Nodes for First Fuzzy Variable

Variation 1

Obj_1	Energy (g_1)	Distance (g_2)	Low Energy	Short Distance
4.950495	0.7920671	2.079213	$e_1 = 10$	$d_1 = 100$
			$e_2 = 16$	$d_2 = 42$
			$e_3 = 14$	$d_3 = 175$
			$e_4 = 29$	$d_4 = 196$
			$e_5 = 20$	$d_5 = 156$

Variation 2

Obj_1	Energy(g_1)	Distance (g_2)	Low Energy	Short Distance
3.2216550	0.6441934	1.675310	$e_1 = 15$	$d_1 = 90$
			$e_2 = 20$	$d_2 = 52$
			$e_3 = 24$	$d_3 = 145$
			$e_4 = 19$	$d_4 = 176$
			$e_5 = 25$	$d_5 = 186$

TABLE 6.5
Dataset for Round 1 Under 100 Sensor Nodes for Second Fuzzy Variable

Variation 1

Obj_2	Energy (g_1)	Distance (g_2)	Medium Energy	Medium Distance
0.2740029	0.1751720	0.4932724	$e_6 = 20$	$d_6 = 290$
			$e_7 = 56$	$d_7 = 250$
			$e_8 = 64$	$d_8 = 180$
			$e_9 = 49$	$d_9 = 280$
			$e_{10} = 40$	$d10 = 316$

Variation 2

Obj_2	Energy(g_1)	Distance (g_2)	Medium Energy	Medium Distance
0.2702703	0.8104624E-01	0.5135190	$e_6 = 30$	$d_6 = 190$
			$e7 = 46$	$d7 = 220$
			$e_8 = 54$	$d_8 = 200$
			$e_9 = 69$	$d_9 = 240$
			$e_{10} = 70$	$d10 = 336$

TABLE 6.6
Dataset for Round 1 Under 100 Sensor Nodes for Third Fuzzy Variable

Variation 1

Obj_3	Energy (g_1)	Distance (g_2)	High Energy	Long Distance
0.6698642E-01	0.5361856E-01	0.2532024	$e_{11} = 65$	$d_{11} = 382$
			$e_{12} = 70$	$d1_2 = 490$
			$e_{13} = 100$	$d1_3 = 469$
			$e_{14} = 92$	$d1_4 = 400$
			$e_{15} = 80$	$d1_5 = 378$

Variation 2

Obj_3	Energy (g_1)	Distance (g_2)	High Energy	Long Distance
0.6944012E-01	0.5204275E-01	0.2583247	$e_{11} = 75$	$d_{11} = 372$
			$e_{12} = 80$	$d_{12} = 470$
			$e_{13} = 89$	$d_{13} = 439$
			$e_{14} = 95$	$d_{14} = 420$
			$e_{15} = 100$	$d_{15} = 478$

TABLE 6.7
Dataset for Round 2 Under 200 Sensor Nodes for First Fuzzy Variable

Variation 1

Obj_4	Energy (g_1)	Distance (g_2)	Low Energy	Short Distance
4.400924	0.5500737	2.024437	$e_1 = 30$	$d_1 = 200$
			$e_2 = 20$	$d_2 = 180$
			$e_3 = 12$	$d_3 = 190$
			$e_4 = 25$	$d_4 = 92$
			$e_5 = 23$	$d_5 = 160$

Variation 2

Obj_4	Energy (g_1)	Distance (g_2)	Low Energy	Short Distance
3.143419	0.3928804	1.728891	$e_1 = 15$	$d_1 = 180$
			$e_2 = 25$	$d_2 = 110$
			$e_3 = 19$	$d_3 = 150$
			$e_4 = 21$	$d_4 = 192$
			$e_5 = 27$	$d_5 = 130$

TABLE 6.8

Dataset for Round 2 Under 200 Sensor Nodes for Second Fuzzy Variable

Variation 1

Obj_5	Energy (g_1)	Distance (g_2)	Medium Energy	Medium Distance
1.089176	0.1361132	1.034722	$e_6 = 25$	$d_6 = 190$
			$e_7 = 35$	$d_7 = 320$
			$e_8 = 42$	$d_8 = 340$
			$e_9 = 52$	$d_9 = 270$
			$e_{10} = 70$	$d_{10} = 225$

Variation 2

Obj_5	Energy (g_1)	Distance (g_2)	Medium Energy	Medium Distance
1.161948	0.2610751	1.0458443	$e_6 = 45$	$d_6 = 180$
			$e_7 = 65$	$d_7 = 318$
			$e_8 = 41$	$d_8 = 280$
			$e_9 = 62$	$d_9 = 310$
			$e_{10} = 70$	$d_{10} = 325$

TABLE 6.9

Dataset for Round 2 Under 200 Sensor Nodes for Third Fuzzy Variable

Variation 1

Obj_6	Energy (g_1)	Distance (g_2)	High Energy	Long Distance
0.2751713	0.1266605	0.5090466	e11 = 65	d11 = 380
			e12 = 75	d12 = 490
			e13 = 82	d13 = 400
			e14 = 92	d14 = 370
			e15 = 99	d15 = 450

Variation 2

Obj_6	Energy (g_1)	Distance (g_2)	High Energy	Long Distance
0.3255500	0.1009007	0.5615773	e11 = 62	d11 = 345
			e12 = 70	d12 = 400
			e13 = 80	d13 = 420
			e14 = 95	d14 = 390
			e15 = 100	d15 = 475

Minimize: $Obj1 = (g_1)^2 + (g_2)^2$

Subject to constraints: $e_1 g_1 + d_1 g_2 \geq 100$

$e_2 g_1 + d_2 g_2 \geq 100$

$e_3 g_1 + d_3 g_2 \geq 100$ (6)

$e_4 g_1 + d_4 g_2 \geq 100$

$e_5 g_1 + d_5 g_2 \geq 100$

Minimize: $\text{Obj2} = (g_1)^2 + (g_2)^2$
Subject to constraints: $e_6 g_1 + d_6 g_2 \geq 100$
$e_7 g_1 + d_7 g_2 \geq 100$
$e_8 g_1 + d_8 g_2 \geq 100$ (7)
$e_9 g_1 + d_9 g_2 \geq 100$
$e_{10} g_1 + d_{10} g_2 \geq 100$

Minimize: $\text{Obj3} = (g_1)^2 + (g_2)^2$
Subject to constraints: $e_{11} g_1 + d_{11} g_2 \geq 100$
$e_{12} g_1 + d_{12} g_2 \geq 100$
$e_{13} g_1 + d_{13} g_2 \geq 100$ (8)
$e_{14} g_1 + d_{14} g_2 \geq 100$
$e_{15} g_1 + d_{15} g_2 \geq 100$

Minimize: $\text{Obj4} = (g_1)^2 + (g_2)^2$
Subject to constraints: $e_1 g_1 + d_1 g_2 \geq 200$
$e_2 g_1 + d_2 g_2 \geq 200$
$e_3 g_1 + d_3 g_2 \geq 200$ (9)
$e_4 g_1 + d_4 g_2 \geq 200$
$e_5 g_1 + d_5 g_2 \geq 200$

Minimize: $\text{Obj5} = (g_1)^2 + (g_2)^2$
Subject to constraints: $e_6 g_1 + d_6 g_2 \geq 200$
$e_7 g_1 + d_7 g_2 \geq 200$
$e_8 g_1 + d_8 g_2 \geq 200$ (10)
$e_9 g_1 + d_9 g_2 \geq 200$
$e_{10} g_1 + d_{10} g_2 \geq 200$

Minimize: $\text{Obj6} = (g_1)^2 + (g_2)^2$
Subject to constraints: $e_{11} g_1 + d_{11} g_2 \geq 200$
$e_{12} g_1 + d_{12} g_2 \geq 200$
$e_{13} g_1 + d_{13} g_2 \geq 200$ (11)
$e_{14} g_1 + d_{14} g_2 \geq 200$
$e_{15} g_1 + d_{15} g_2 \geq 200$

6.4 SIMULATION AND ANALYSIS

The suggested technique is simulated and validated using the LINGO optimization simulator, which is used to optimize nonlinear objectives and constraints. Six
total nonlinear goal functions with linear constraints are employed in this research.
There are five linear constraints in each objective function. So here, total constrains
is 6 × 5, i.e., 30. In this simulation, nodes used are 100 to 200, based on two variations, namely, variation 1 and variation 2. Total fuzzy linguistic variables used is 6,
3 for energy consumption, namely, low, medium, and high, and 3 for distance, such
namely, short, medium, and long. Energy used is 100 J, and distance used is 500
meters. A summary of the simulation parameters is shown in Table 6.10.

TABLE 6.10
Simulation Parameters

Parameter for simulation	Description
Optimization software	LINGO
Total objective function	6
Constraints	30
Number of nodes	100–200
Number of variations	2
Number of rounds	12
Number of linguistic variables	2×3
Network parameters	Energy consumption, distance
Energy	100 J
Distance	500 meters

Figures 6.1 to 6.3 show the output of nonlinear formulation of quadratic programming based on the first, second, and third linguistic variables of energy and distance. These illustrations are under 100 nodes for "variation 1." Here, linguistic variables increase in a chronological order based on fuzzy nature. In Figure 1, the linguistic variable is low and short for input parameters, so here the value of the objective function is 4.950495 by decision variables 0.7920671 and 2.079213. In Figure 6.2, the linguistic variables are medium and medium for the input parameters, so here the value of the objective function is 0.2740029 by decision variables 0.1751720 and 0.4932724. In Figure 6.3, the linguistic variables are high and long for the input parameters, so here the value of the objective function is 0.6698642E-01 by decision variables 0.5361856E-01 and 0.2532024. Hence, it is observed that, when linguistic variables are increasing in nature, their optimized values decrease.

Figures 6.4 to 6.6 show the output of nonlinear formulation of quadratic programming based on the first, second, and third linguistic variables of energy and distance. These illustrations are under 200 nodes for "variation 1." Here, the linguistic variables increase chronologically based on fuzzy nature. In Figure 6.4, the linguistic variables are low and short for the input parameters, so here the value of the objective function is 4.400924 by decision variables 0.5500737 and 2.024437. In Figure 6.5, the linguistic variables are medium and medium for the input parameters, so here the value of the objective function is 1.089176 by decision variables 0.1361132 and 1.034722. In Figure 6.6, the linguistic variables are high and long for the input parameter, so here the value of the objective function is 0.2751713 by decision variables 0.1266605 and 0.5090466. Hence, it is observed that, when linguistic variables are increasing in nature, their optimized values decrease. In Figures 6.4 to 6.6, the number of nodes increases to 200, but the nature is the same as in the previous illustrations.

Figures 6.7 to 6.9 show the output of nonlinear formulation of quadratic programming based on the first, second, and third linguistic variables of energy and distance.

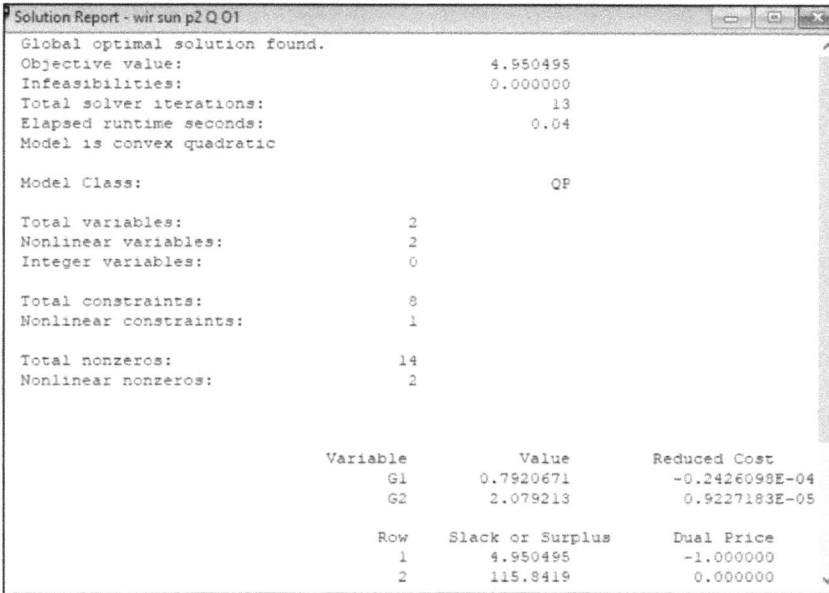

FIGURE 6.1 Variation 1 for 100 nodes with first linguistic variable of input parameters.

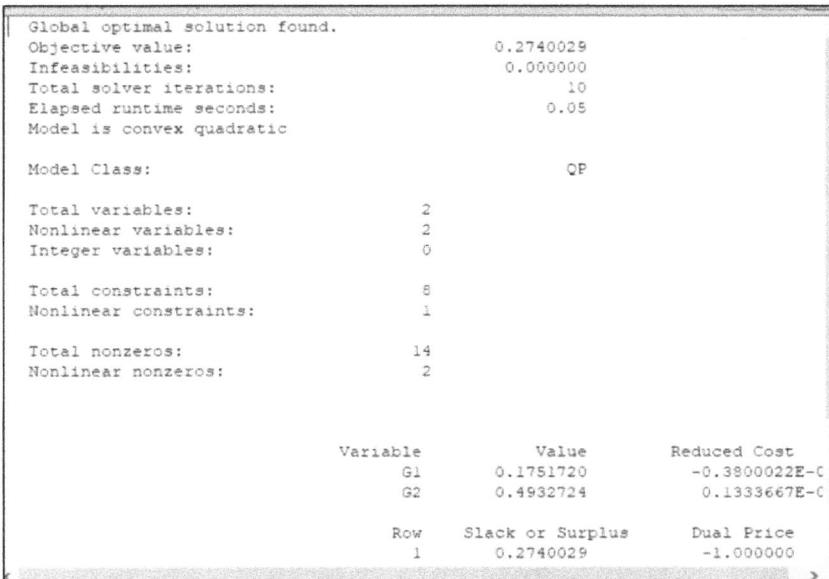

FIGURE 6.2 Variation 1 for 100 nodes with second linguistic variable of input parameters.

```
Global optimal solution found.
Objective value:                              0.6698642E-01
Infeasibilities:                              0.000000
Total solver iterations:                            10
Elapsed runtime seconds:                          0.06
Model is convex quadratic

Model Class:                                        QP

Total variables:                     2
Nonlinear variables:                 2
Integer variables:                   0

Total constraints:                   8
Nonlinear constraints:               1

Total nonzeros:                     14
Nonlinear nonzeros:                  2


                        Variable           Value        Reduced Cost
                              G1      0.5361856E-01      0.5968841E-04
                              G2        0.2532024       -0.1278010E-04

                             Row   Slack or Surplus      Dual Price
                               1      0.6698642E-01       -1.000000
                               2        0.2085356       -0.5769022E-07
```

FIGURE 6.3   Variation 1 for 100 nodes with third linguistic variable of input parameters.

```
Solution Report - wir sun p2 Q Q4                                    _ □ ▬
Global optimal solution found.
Objective value:                              4.400924
Infeasibilities:                              0.000000
Total solver iterations:                            12
Elapsed runtime seconds:                          0.04
Model is convex quadratic

Model Class:                                        QP

Total variables:                     2
Nonlinear variables:                 2
Integer variables:                   0

Total constraints:                   8
Nonlinear constraints:               1

Total nonzeros:                     14
Nonlinear nonzeros:                  2


                        Variable           Value        Reduced Cost
                              G1        0.5500737       -0.8370017E-04
                              G2         2.024437        0.2265881E-04

                             Row   Slack or Surplus      Dual Price
                               1        4.400924         -1.000000
                               2         221.3895         0.000000
```

FIGURE 6.4   Variation 1 for 200 nodes with first linguistic variable of input parameters.

```
Solution Report - wir sun p2 Q 05

Global optimal solution found.
Objective value:                         1.089176
Infeasibilities:                         0.000000
Total solver iterations:                       11
Elapsed runtime seconds:                     0.04
Model is convex quadratic

Model Class:                                   QP

Total variables:                 2
Nonlinear variables:             2
Integer variables:               0

Total constraints:               8
Nonlinear constraints:           1

Total nonzeros:                 14
Nonlinear nonzeros:              2

               Variable           Value        Reduced Cost
                     G1       0.1361132      -0.6765914E-04
                     G2        1.034722       0.8701987E-05

                    Row   Slack or Surplus        Dual Price
                      1        1.089176           -1.000000
                      2    0.5081752E-05      -0.1089176E-01
                      3        135.8750            0.000000
                      4        157.5222            0.000000
                      5        86.45282            0.000000
                      6        42.34037            0.000000
                      7       0.1361132            0.000000
                      8        1.034722            0.000000
```

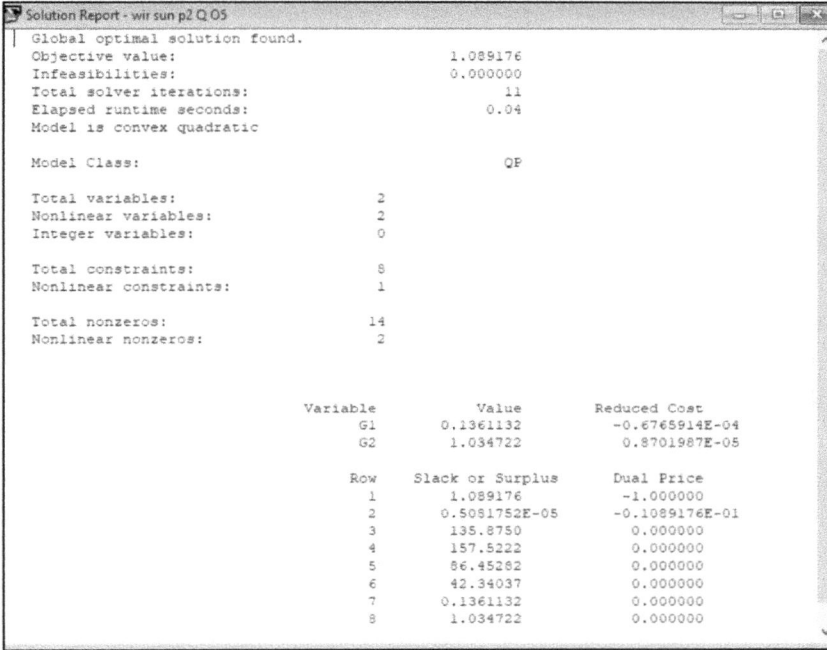

FIGURE 6.5   Variation 1 for 200 nodes with second linguistic variable of input parameters.

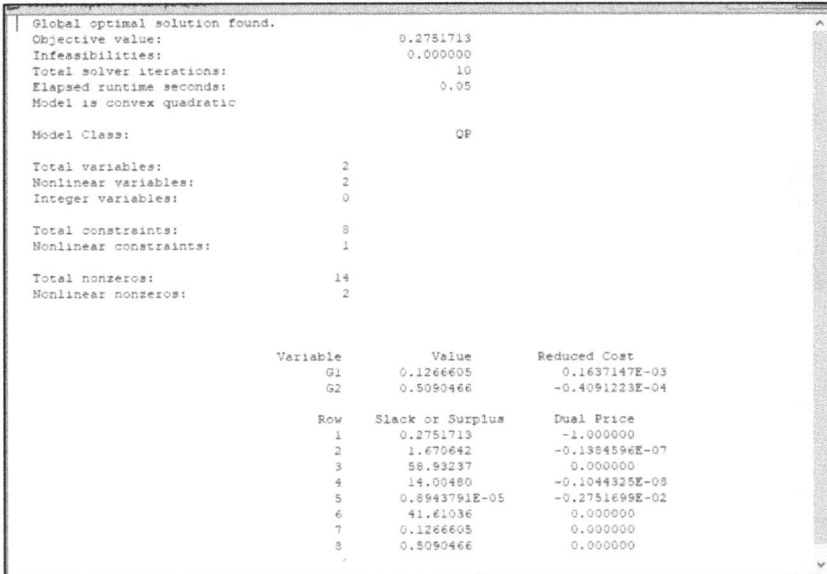

```
Global optimal solution found.
Objective value:                         0.2751713
Infeasibilities:                         0.000000
Total solver iterations:                       10
Elapsed runtime seconds:                     0.05
Model is convex quadratic

Model Class:                                   QP

Total variables:                 2
Nonlinear variables:             2
Integer variables:               0

Total constraints:               8
Nonlinear constraints:           1

Total nonzeros:                 14
Nonlinear nonzeros:              2

               Variable           Value        Reduced Cost
                     G1       0.1266605       0.1637147E-03
                     G2       0.5090466      -0.4091223E-04

                    Row   Slack or Surplus        Dual Price
                      1       0.2751713           -1.000000
                      2        1.670642      -0.1384596E-07
                      3        58.93237            0.000000
                      4        14.00480      -0.1044325E-08
                      5    0.8943791E-05      -0.2751699E-02
                      6        41.61036            0.000000
                      7       0.1266605            0.000000
                      8       0.5090466            0.000000
```

FIGURE 6.6   Variation 1 for 200 nodes with third linguistic variable of input parameters.

```
Solution Report - wir sun p2 Q v21                                          - □ X
  Global optimal solution found.
  Objective value:                                    3.221650
  Infeasibilities:                                    0.000000
  Total solver iterations:                                  12
  Elapsed runtime seconds:                                0.04
  Model is convex quadratic

  Model Class:                                              QP

  Total variables:                      2
  Nonlinear variables:                  2
  Integer variables:                    0

  Total constraints:                    8
  Nonlinear constraints:                1

  Total nonzeros:                      14
  Nonlinear nonzeros:                   2


              Variable            Value         Reduced Cost
                    G1         0.6441934        -0.2734699E-03
                    G2         1.675310          0.1046249E-03

                   Row  Slack or Surplus           Dual Price
                     1         3.221650          -1.000000
                     2        60.44083            0.000000
                     3         0.6575197E-05     -0.6443301E-01
                     4       158.3806             0.000000
                     5       207.0943             0.000000
                     6       227.7126             0.1063586E-08
                     7         0.6441934          0.000000
                     8         1.675310           0.000000
```

FIGURE 6.7   Variation 2 for 100 nodes with first linguistic variable of input parameters.

```
Solution Report - wir sun p2 Q v22                                          - □ X
  Global optimal solution found.
  Objective value:                                    0.2702703
  Infeasibilities:                                    0.000000
  Total solver iterations:                                  10
  Elapsed runtime seconds:                                0.04
  Model is convex quadratic

  Model Class:                                              QP

  Total variables:                      2
  Nonlinear variables:                  2
  Integer variables:                    0

  Total constraints:                    8
  Nonlinear constraints:                1

  Total nonzeros:                      14
  Nonlinear nonzeros:                   2


              Variable            Value         Reduced Cost
                    G1         0.8104624E-01    -0.6971891E-04
                    G2         0.5135190         0.1094153E-04

                   Row  Slack or Surplus           Dual Price
                     1         0.2702703         -1.000000
                     2         0.3854529E-06     -0.5405405E-02
                     3        16.70231            0.000000
                     4         7.080300           0.000000
                     5        28.83675            0.000000
                     6        78.21563            0.000000
                     7         0.8104624E-01      0.000000
                     8         0.5135190          0.000000
```

FIGURE 6.8   Variation 2 for 100 nodes with second linguistic variable of input parameters.

These illustrations are under 100 nodes for "variation 2." Here, the linguistic variables increase chronologically based on fuzzy nature. In Figure 6.7, the linguistic variables are low and short for the input parameters, so here the value of the objective function is 3.2216550 by decision variables 0.6441934 and 1.675310. In Figure 6.8, the linguistic variables are medium and medium for the input parameters, so here the value of the objective function is 0.2702703 by decision variables 0.8104624E-01 and 0.5135190. In Figure 6.9, the linguistic variables are high and long for the input parameters, so here the value of the objective function is 0.6944012E-01 by decision variables 0.5204275E-01 and 0.2583247. Hence, it is observed, when linguistic variables are increasing in nature, their optimized values decrease.

Figures 6.10 to 6.12 show the output of nonlinear formulation of quadratic programming based on the first, second, and third linguistic variables of energy and distance. These illustrations are under 200 nodes for "variation 2." Here, the linguistic variables increase chronologically based on fuzzy nature. In Figure 10, the linguistic variables are low and short for the input parameters, so here the value of the objective function is 3.143419 by decision variables 0.3928804 and 1.728891. In Figure 6.11, the linguistic variables are medium and medium for the input parameters, so here the value of the objective function is 1.161948 by decision variables 0.2610751 and 1.045843. In Figure 6.12, the linguistic variables are high and long for the input parameters, so here the value of the objective function is 0.3255500 by decision variables 0.1009007 and 0.5615773. Hence, it is observed that, when linguistic variables are increasing in nature, their optimized values decrease. In Figures 6.10 to 6.12, the number of nodes increases to 200, but the nature is the same as in the previous illustrations.

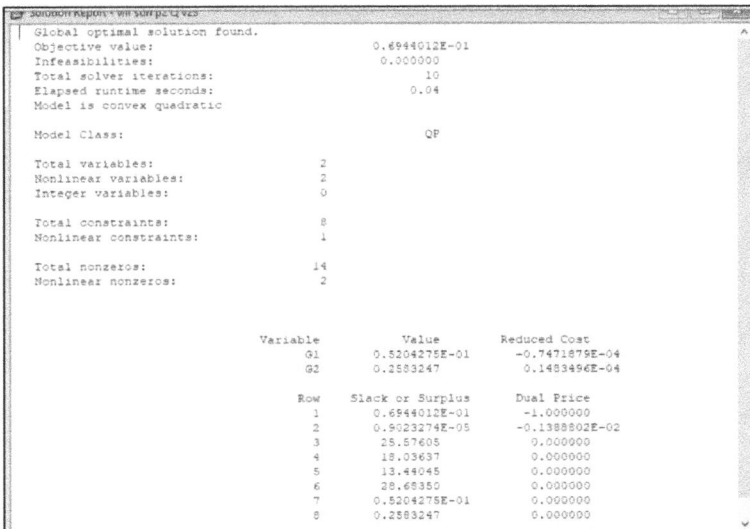

FIGURE 6.9 Variation 2 for 100 nodes with third linguistic variable of input parameters.

```
Solution Report - wir sun p2 Q v24
  Global optimal solution found.
  Objective value:                        3.143419
  Infeasibilities:                        0.000000
  Total solver iterations:                      10
  Elapsed runtime seconds:                    0.04
  Model is convex quadratic

  Model Class:                                  QP

  Total variables:              2
  Nonlinear variables:          2
  Integer variables:            0

  Total constraints:            8
  Nonlinear constraints:        1

  Total nonzeros:              14
  Nonlinear nonzeros:           2


                     Variable          Value        Reduced Cost
                           G1      0.3928804       -0.9392961E-04
                           G2       1.728891        0.2119624E-04

                          Row   Slack or Surplus        Dual Price
                            1       3.143419          -1.000000
                            2       117.0936           0.000000
                            3      0.1554560E-05      -0.3143418E-01
                            4       66.79835           0.000000
                            5       140.1975           0.000000
                            6       35.36358          -0.1883601E-08
                            7      0.3928804           0.000000
                            8       1.728891           0.000000
```

FIGURE 6.10 Variation 2 for 200 nodes with first linguistic variable of input parameters.

```
Solution Report - wir sun p2 Q v25
  Global optimal solution found.
  Objective value:                        1.161948
  Infeasibilities:                        0.000000
  Total solver iterations:                       9
  Elapsed runtime seconds:                    0.04
  Model is convex quadratic

  Model Class:                                  QP

  Total variables:              2
  Nonlinear variables:          2
  Integer variables:            0

  Total constraints:            8
  Nonlinear constraints:        1

  Total nonzeros:              14
  Nonlinear nonzeros:           2


                     Variable          Value        Reduced Cost
                           G1      0.2610751       -0.7268683E-03
                           G2       1.045843        0.1785969E-03

                          Row   Slack or Surplus        Dual Price
                            1       1.161948          -1.000000
                            2      0.1502812E-03      -0.1161950E-01
                            3       149.5480           0.2788510E-08
                            4       103.8402          -0.1942500E-08
                            5       140.3980           0.2463378E-08
                            6       158.1743           0.3129008E-08
                            7      0.2610751           0.000000
                            8       1.045843           0.000000
```

FIGURE 6.11 Variation 2 for 200 nodes with second linguistic variable of input parameters.

```
Solution Report - wir sun p2 U v2b                                    ☐ ☐ ☐
  Global optimal solution found.
  Objective value:                          0.3255500
  Infeasibilities:                          0.000000
  Total solver iterations:                        11
  Elapsed runtime seconds:                    0.04
  Model is convex quadratic

  Model Class:                                   QP

  Total variables:                  2
  Nonlinear variables:              2
  Integer variables:                0

  Total constraints:                8
  Nonlinear constraints:            1

  Total nonzeros:                  14
  Nonlinear nonzeros:               2


                        Variable          Value        Reduced Cost
                            G1        0.1009007      -0.3964777E-04
                            G2        0.5615773       0.7065469E-05

                        Row      Slack or Surplus       Dual Price
                          1        0.3255500          -1.000000
                          2        0.1953675E-05      -0.3255500E-02
                          3        31.69396            0.000000
                          4        43.93451            0.000000
                          5        28.60070            0.000000
                          6        76.83927            0.000000
                          7        0.1009007           0.000000
                          8        0.5615773           0.000000
```

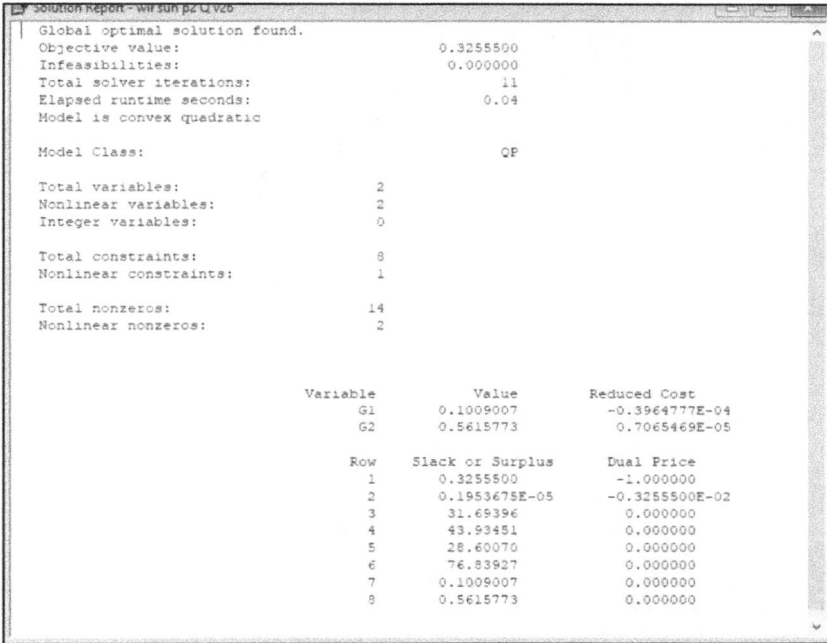

FIGURE 6.12   Variation 2 for 200 nodes with third linguistic variable of input parameters.

6.5   CONCLUSION

The current paper is based on the optimization of energy and distance based on network lifetime optimization. This optimization is based on quadratic programming for the purpose of WSN parameter handling. The work is formulated with fuzzy logic for the purpose of optimization based on different network parameters. It helps optimize several information based on certain information that help affect network information based on network lifetime. It helps optimize several information based on fuzzy linguistic variable that help model applications efficiently and effectively.

6.6   REFERENCES

[1] Mishra, B. P., Panigrahi, T., Wilson, A. M., & Sabat, S. L. (2022). A robust diffusion algorithm using logarithmic hyperbolic cosine cost function for channel estimation in wireless sensor network under impulsive noise environment. *Digital Signal Processing*, 103384, https://doi.org/10.1016/j.dsp.2022.103384

[2] Mahalle, P. N., Shelar, P. A., Shinde, G. R., & Dey, N. (2021). Introduction to underwater wireless sensor networks. In *The Underwater World for Digital Data Transmission* (pp. 1–21). Springer, Singapore.

[3] Toklu, S., & Erdem, O. A. (2014). BSC-MAC: Energy efficiency in wireless sensor networks with base station control. *Computer Networks*, 59, 91–100.

[4] Moh'd Alia, O. (2017). Dynamic relocation of mobile base station in wireless sensor networks using a cluster-based harmony search algorithm. *Information Sciences*, 385, 76–95.

[5] Das, S. K., Maheswari, V., & Sharma, A. (2021). Wireless networks: Applications, challenges, and security issues. In *Architectural Wireless Networks Solutions and Security Issues* (pp. 1–10). Springer, Singapore.

[6] Dey, N., Hassanien, A. E., Bhatt, C., Ashour, A., & Satapathy, S. C. (Eds.). (2018). *Internet of Things and Big Data Analytics toward Next-generation Intelligence* (Vol. 35). Springer, Berlin.

[7] Singh, A. K., & Pamula, R. (2021). An efficient and intelligent routing strategy for vehicular delay tolerant networks. *Wireless Networks*, 27(1), 383–400.

[8] Dey, N., Samanta, S., Chakraborty, S., Das, A., Chaudhuri, S. S., & Suri, J. S. (2014). Firefly algorithm for optimization of scaling factors during embedding of manifold medical information: An application in ophthalmology imaging. *Journal of Medical Imaging and Health Informatics*, 4(3), 384–394.

[9] Keerthika, M., & Shanmugapriya, D. (2021). Wireless sensor networks: Active and passive attacks-vulnerabilities and countermeasures. *Global Transitions Proceedings*, 2(2), 362–367.

[10] Das, S. K., Dao, T. P., & Perumal, T. (Eds.). (2021). *Nature-Inspired Computing for Smart Application Design*. Springer Nature, Singapore.

[11] Prasad, R. (2021). Enhanced energy efficient secure routing protocol for mobile ad-hoc network. *Global Transitions Proceedings*, https://doi.org/10.1016/j.gltp.2021.10.001.

[12] Wan, J., & Chen, J. (2022). AHP based relay selection strategy for energy harvesting wireless sensor networks. *Future Generation Computer Systems*, 128, 36–44.

[13] Singh, N. C., & Sharma, A. (2020). Resilience of mobile ad hoc networks to security attacks and optimization of routing process. *Materials Today: Proceedings*, https://doi.org/10.1016/j.matpr.2020.09.622.

[14] Das, S. K. (2021). Smart design and its applications: Challenges and techniques. *Nature-Inspired Computing for Smart Application Design*, 1.

[15] Misra, Y., Krishnaveni, K., & Rajasekaran, A. S. (2022). Implementation of NLOS based FPGA for distance estimation of elderly indoor wireless sensor networks. *Materials Today: Proceedings*, https://doi.org/10.1016/j.matpr.2022.01.087.

[16] Anand, M., Balaji, N., Bharathiraja, N., & Antonidoss, A. (2021). A controlled framework for reliable multicast routing protocol in mobile ad hoc network. *Materials Today: Proceedings*, https://doi.org/10.1016/j.matpr.2020.10.902.

[17] Dey, N., Ashour, A. S., Shi, F., Fong, S. J., & Sherratt, R. S. (2017). Developing residential wireless sensor networks for ECG healthcare monitoring. *IEEE Transactions on Consumer Electronics*, 63(4), 442–449.

[18] Wang, F., & Hu, H. (2021). Coverage hole detection method of wireless sensor network based on clustering algorithm. *Measurement*, 179, 109449, https://doi.org/10.1016/j.measurement.2021.109449.

[19] Tahir, A., Shah, N., Abid, S. A., Khan, W. Z., Bashir, A. K., & Zikria, Y. B. (2021). A three-dimensional clustered peer-to-peer overlay protocol for mobile ad hoc networks. *Computers & Electrical Engineering*, 94, 107364, https://doi.org/10.1016/j.compeleceng.2021.107364.

[20] Temene, N., Sergiou, C., Georgiou, C., & Vassiliou, V. (2022). A survey on mobility in wireless sensor networks. *Ad Hoc Networks*, 125, 102726, https://doi.org/10.1016/j.adhoc.2021.102726.

[21] De, D., Mukherjee, A., Das, S. K., & Dey, N. (2020). Wireless sensor network: Applications, challenges, and algorithms. In *Nature Inspired Computing for Wireless Sensor Networks* (pp. 1–18). Springer, Singapore.

[22] Sharma, A. S., & Kim, D. S. (2021). Energy efficient multipath ant colony based routing algorithm for mobile ad hoc networks. *Ad Hoc Networks*, 113, 102396, https://doi.org/10.1016/j.adhoc.2020.102396.

[23] Yousefpoor, M. S., Yousefpoor, E., Barati, H., Barati, A., Movaghar, A., & Hosseinzadeh, M. (2021). Secure data aggregation methods and countermeasures against various attacks in wireless sensor networks: A comprehensive review. *Journal of Network and Computer Applications*, 103118, https://doi.org/10.1016/j.jnca.2021.103118.

[24] Das, S. K., Samanta, S., Dey, N., Patel, B. S., & Hassanien, A. E. (Eds.). (2021). *Architectural Wireless Networks Solutions and Security Issues*. Springer, Singapore.

[25] Lee, S., Youn, J., & Jung, B. C. (2020). A cooperative phase-steering technique in spectrum sharing-based military mobile ad hoc networks. *ICT Express*, 6(2), 83–86.

[26] Singh, A. K., Pamula, R., Jain, P. K., & Srivastava, G. (2021). An efficient vehicular-relay selection scheme for vehicular communication. *Soft Computing*, 1–17.

[27] De, D., Mukherjee, A., Das, S. K., & Dey, N. (Eds.). (2020). *Nature Inspired Computing for Wireless Sensor Networks*. Springer, Singapore.

[28] Ahmed, H. I., Nasr, A. A., Abdel-Mageid, S. M., & Aslan, H. K. (2021). DADEM: Distributed attack detection model based on big data analytics for the enhancement of the security of internet of things (IoT). *International Journal of Ambient Computing and Intelligence (IJACI)*, 12(1), 114–139.

[29] Sholla, S., Mir, R. N., & Chishti, M. A. (2021). A fuzzy logic-based method for incorporating ethics in the internet of things. *International Journal of Ambient Computing and Intelligence (IJACI)*, 12(3), 98–122.

[30] Balusa, B. C., & Gorai, A. K. (2021). Development of fuzzy pattern recognition model for underground metal mining method selection. *International Journal of Ambient Computing and Intelligence (IJACI)*, 12(4), 64–78.

Section 3

Data Analysis and Prediction

7 A Review Based on Prediction Analysis to Mitigate the Issues of COVID-19

Santosh Kumar Das and Aditya Sharma

CONTENTS

7.1 Introduction 117
7.2 Literature Review 118
7.3 Artificial Intelligence to Mitigate and Analyze the Disease 127
7.4 Conclusions..130
7.5 References..130

7.1 INTRODUCTION

In the last few years, several types of diseases have been raised due to the COVID-19 pandemic. Initially, it started with a few complications; after that, it has changed frequently based on several mutations of the virus [1–2]. Several works have been proposed based on artificial intelligence techniques for the purpose of solving issues related to the COVID-19 pandemic. Each of these issues is based on some analysis and prediction that require several types of diagnosis. Several works have been proposed using artificial intelligence techniques to solve several issues. In [3–4], the authors illustrated several works based on machine learning algorithm, smart application design, wireless network, and wireless sensor networks. The stated techniques help maintain and analyze the issue to determine solutions on any area or domain. Due to this pandemic, several issues have been raised in the sector of education, business, finance, e-commerce, media, banking, etc. Each of the sectors may be government or private or semigovernment or undertaking, among others. Each sector is affected fully or partially due to this pandemic, but its effect on all sectors is shown in terms of diseases that are reported in the hospital.

The proposed article is a review-based article on the fusion of several papers that help in modeling solutions to several issues and problems with their parameters in a single framework. This framework of the paper helps guide several readers and researchers toward the analysis of different diseases related to COVID-19.

It also shows the different techniques used in the last few years for resolving COVID issues under the domain of artificial intelligence. It illustrates the different types of algorithms used in this pandemic and also discusses the working principles and nature of artificial intelligence based on some phases for modeling purpose. This modeling helps resolve the issues efficiently to make proper solution.

The emainning part of this article is divided as thus: Section 2 describes a detailed analysis of literature review regarding COVID-19 based on different techniques on artificial intelligence. Section 3 discusses the methodologies used to mitigate and deal with several issues raised by diseases related to COVID-19. Section 4 is the conclusion.

7.2 LITERATURE REVIEW

In the last few years, several research related to COVID-19 have been proposed using artificial intelligence techniques to mitigate the issues of the disease. In this section, some works have been discussed, as Alhasan and Hasaneen [5] designed a method based on a digital image system based on artificial intelligence techniques related to COVID-19. The method also uses the technique of natural language process for the purpose of analyzing data. It is based on a public health analysis system that uses the computed tomography system. It is based on contact tracing and its analysis system based on an enabling system. It helps process mobile CT images for the analysis of different information. The work is based on stream innovation and an analysis system that helps in COVID-19 analysis and prediction. The work is the fusion of artificial intelligence and machine learning system for enhancing the system with model analysis. Lan et al. [6] designed a method for patient analysis and its management for the purpose of COVID-19. The method is based on an artificial intelligence system for the analysis of the coronavirus. The method is based on a public health analysis system for the purpose of controlling and managing different forecasting information. This management also uses several types of vaccines and drugs to manage the disease. It helps reduce several challenges and accelerate different types of achievements. Tarik et al. [7] designed a method of student performance prediction and analysis. This prediction is based on COVID-19, which helps in managing several issues during lockdown and conducting classes from home. The method is based on artificial intelligence, and the analysis is done among students of Moroccan descent. This is one of the regions under Guelmim Oued Noun. Some of these issues are based on nature-inspired computing for wireless networks [8]. Haleem et al. [9] designed a cardiology analysis and prediction for COVID-19. The work is based on an artificial intelligence technique. In this work, several information are accessed based on the pandemic period from health-care systems. It is based on the cardiology treatment and its analysis on COVID-19. It helps predict and analyze several prevention techniques and the interaction of information. It also helps analyze several information based on different innovation systems. It is based on a critical heart surgery system. The method uses several techniques of artificial intelligence, such as support vector machine and artificial neural network. Zhang et al. [10] designed a validation system for managing different factors of prognosis for handling different patients during the COVID-19 pandemic. The study is based on retrospective analysis based on an artificial intelligence technique. The analysis technique of this method is based on the coronavirus disease. This virus is based on a novel virus system on global pandemic analysis. Several techniques are used in this model for the purpose of artificial intelligence, such as artificial neural network and the Lasso operator, which is based on neural network. Born et al. [11] designed an analysis model based on medical image processing for COVID-19. This work is based on a medical image survey report on clinical image analysis. It is based on several relevance system of images for analysis of vast recommendations. The work is also based on a systematic review of employee details and performance system along with patient satisfaction. It is based on several deployment and practices of clinical diagnosis. This modeling helps model several wireless networks and their related formulation [12]. Barnawi et al. [13] designed a method for COVID-19 screening and analysis related to the pandemic. It uses a thermal imaging system that uses a fusion of artificial intelligence and internet of things. The method uses several metrics for the purpose of evaluation. It uses several sensors for analysis based on the classifier method of machine leaning with the fusion of artificial intelligence. Cheema et al. [14] designed a method related to COVID-19. In this method, the author needs to analyze several information based on medical analysis of POCUS in the ICU. In this work, the model is based on point-of-care ultrasound analysis and its management. It uses artificial intelligence to enable the system and its services. In this method, cardiac ultrasound is used for medical diagnosis. Sucharitha and Chary [15] designed a prediction effect of COVID-19 based on several variations and analyses. It is based on several facilities and striving system based on different collaborations. The work is based on cognitive system and different integration systems. It is based on real-time analysis and its management, which helps in depth exploratory data analysis system of management. Several analyses and features are used in this system for handling global datasets, with several predicting analysis and management. The work is based on a biomedical analysis system and its management for handling several information based on disease analysis. It helps in modeling several problems also for the application of nature optimization [16]. Jiao et al. [17] designed a patient analysis system based on the prognostication system for the purpose of X-ray analysis of chest based on different clinical data and information. It is based on retrospective analysis of different studies and disease management. The work is based on an artificial intelligence system and its different innovation techniques. It helps manage and develop several clinical data and information based on different progression and its information system. The model is trained based on deep neural network analysis for the purpose of feature extraction and risk modeling system. Karaman et al. [18] designed a method of social distancing analysis with camera analysis for whether social distancing is maintained properly or not. The work is based on an artificial intelligence technique on network analysis systems and its predictions. The main purpose of this camera is for the intrusion-detection system of different types of analysis. Social distancing is maintained by convolution neural network and its analysis parameters. It uses the Raspberry Pi software for the camera and helps in different types of prediction systems. It helps in integrated system analysis to determine the violation system of distance. Madani et al. [19] proposed a method of fake news detection. It is based on Moroccan tweet analysis. In this model, several data are traveled and analyzed based on world information systems. The work is based on several types of experimental analysis systems for detecting several types of fake news. Fake news creates several types of uncertainties and ambiguities for handling several issues. It promotes lack information of the model for the purpose of system analysis. It helps model several applications to resolve issues on smart applications [20]. Maille et al. [21] designed a model based on smartwatch electrocardiogram to access cardiac rhythm safety analysis. It is based on several types of drug therapy analysis on COVID-19. It helps in managing several issues of the COVID-19 pandemic. The work is based on an artificial intelligence technique for the purpose of QT-logs information and the study system. It helps in managing several types of issues based on home and other monitoring systems. It is based on ECG analysis of different variations. Yaşar et al. [22] designed a method of data and information prediction and severity analysis based on results analysis of different protein system. Severity is based on COVID-19 based on different profile management systems using artificial intelligence techniques. The resulting analysis is based on gradient-boosted tree, or GBT, with a combination of random forest analysis. It is based on several predictive models and analysis systems for effective management. Mohammad-H. Tayarani N. [23] designed a method of battling against several information and parameters of the COVID-19. The work is based on a literature review system on the application of artificial intelligence. This review deals with several information, such as different identification, analysis, patient analysis and monitoring, different types of test analysis, symptom analysis and monitoring, and several types of image analysis and testing. Each of the information and analysis is based on an artificial intelligence technique of pandemic analysis and management. Some of the applications also modeled based on underwater sensor application and management [24]. Zhu et al. [25] designed an illustration of emergency service analysis and management for handling the public health system. It is based on several types of preventions and analyses based on different types of management. The prevention controlling system is based on an example of the country China. It checks several types of stability and harmony management systems. It is based on technical assistance and analysis for helping several emergencies make a decision-making system. Mohammed A.A.Al-qaness et al. [26] designed a model based on forecasting analysis. It is based on outbreak analysis of COVID-19 from information on two countries, Brazil and Russia. In this method, several phenomena are discussed, such as the neurological system, respiratory system, and gastrointestinal system, based on different hazards to the system. Several intelligent techniques are used in this model, such as marine predator algorithm and adaptive neuro fuzzy system based on inference modeling and its enhancement. Finally, this model is analyzed and designed for handling several types of predator systems to outperform the results. Adamidi et al. [27] designed a clinical care system and model related to COVID-19. It is based on pandemic analysis and management. The article is a systematic review of the system and model. In this article, several information is detected based on information from COVID-19. The model is based on an assessment of the severity and predicting system based on the nature of heterogeneous system and modeling. It helps in several types of prediction based on a multimodal system with severity assessment and analysis. It also helps in performance analysis and management. Suri et al. [28] proposed a method of reviewing the characteristic system based on distress syndrome of COVID-19. It is based on an infected lung system based on artificial intelligence techniques. It incorporates several information based on acute respiratory analysis based on syndrome analysis of COVID-19. It deals with several information on pandemic analysis based on technique and information of management. It uses X-ray with several types of CT scan systems with different types of diagnoses and modeling for resolving some challenges within the context of sensor network and application management [29]. Li et al. [30] designed a user study system and its modeling for lung disease analysis based on different types of severity on chest radiograph analysis and its management. The method is based on a multiradiologist system based on techniques of artificial intelligence. It is based on retrospective analysis of a multiradiologist system that helps evaluate several factors of the disease. It helps analyze several interrater agreements based on its death analysis and management. It also uses CXR interpretation and analysis for handling diseases efficiently. Yadav et al. [31] designed a biomarker-based analysis of immunosensors based on medical data analysis. It helps in managing several information based on the internet of medical things information management with artificial intelligence. It is based on COVID-19 disease analysis of different types of diagnoses and management systems. The combination of artificial intelligence and the internet of things on medical information helps in managing several medical information efficiently and effectively. Kumar et al. [32] designed an illustration based on COVID-19 vaccine distribution and analysis based on different factors and information management. The work of this paper is based on the fusion of two techniques, namely, the internet of things and artificial intelligence. The work is based on artificial intelligence with real-time analysis of the health-care system and its methodologies. It incorporates several information, such as access questions and logistic systems, based on real-time applications. Moezzi et al. [33] designed a diagnosis accuracy system based on CT imaging system for the meta-analysis and analysis of different diseases based on COVID-19. The article is based on several types of reviews and analysis on metainformation. The system is based on deep learning with machine learning for sensitivity analysis in predicting and analyzing several symptoms of the disease. Shaikh et al. [34] designed an illustration of clinical analysis and management for different outcomes based on computer tomography analysis. It is based on positron emission and its analysis. The method is based on structural assessment and its analysis. It helps in managing and utilizing several information based on imaging and its analysis. It helps utilize several images based on different predictive power management techniques. Information from this system is based on a prognostication system on modalities of image analysis. Chonde et al. [35] designed a method related to COVID-19 based on power management and analysis of different patients. It is based on English proficiency in the time of COVID-19. The system uses a multilingual care system based on different interpretations. It helps in managing patient conditions based on variations of analysis for the purpose of throughput management. It helps model optimization just like nature-inspired optimization in dealing with several information [36]. Kabra and Singh [37] designed a method of discovery peptide based on the fusion of artificial intelligence and evolutionary algorithm. This is based on COVID-19 disease and its different information. It uses therapeutics for analyzing and predicting several viral sequence information. The main purpose of this system is the analysis of an effective solution based on selectivity and stability. T. Lampejo [38] designed a method for pneumonia analysis with consideration of different types of analysis for the diagnosis of COVID. The work is based on radiological diagnosis for investigation purposes. The article is based on coronavirus 2 analysis and helps analyze how this virus affects the society. In the investigation, information is taken from several hospitals for the purpose of patient and disease diagnoses. Wang et al. [39] designed a method of risk analysis and stratification for patient analysis in the hospital during the COVID-19 pandemic. It is based on artificial intelligence for the echocardiography technique. It helps in managing several prognosis information of COVID-19 that help in several measurement and adjustment for different information. The work is based on different types of clinical tools based on hospital and patient information in managing different systems. Soltan et al. [40] designed a clinical data for the purpose of patient analysis and different types of validation systems for handling information with the help of artificial intelligence. Several information are analyzed and dealt with based on factors of management. It helps in health-care data analysis. It uses several parameters, such as blood test, blood gas, and different types of vital information. Ramella et al. [41] designed a method of radiation analysis for pneumonitis for handling several types of issues based on COVID-19. The work is based on several types of diagnoses and information on radiation pneumonitis for handling lung radiation analysis. The work is based on a statistical analysis for performing different types of roles. It is based on patient analysis for a combination of classification management. Tiwari et al. [42] designed a switching system based on uncertainty management for handling several types of revolutions. It is based on industrial revolution for pandemic handling. It deals with several types of analyses and with management for handling different types of negative analyses. It uses emission trading analysis based on asymmetric analysis on cryptography. It deals with several types of policy uncertainty management on different variations for equity analysis. Haimed et al. [43] designed a method of reverse engineering of COVID-19 analysis and management. It is based on long short-term analysis and management. The method is based on big data with the fusion of artificial intelligence. It helps in several types of evolutions based on the closest analysis of the industry. It helps in managing several information on different factors and parameters. J. J. Hsieh [44] designed a survey issue on artificial intelligence based on different factors. In this system, information are dealt with for diagnosing and analyzing. Each information is based on a diagnosis and factor analysis of several potential information that help in managing several systems. It helps in managing several diagnostic systems based on different parameters of the hospital and the patients. T. L. D. Health [45] designed an artificial intelligence method of savior analysis. It uses several information for digital health analysis. It uses several algorithms of diagnosis and patient analysis based on clinical information. It helps in managing several drug diagnoses on emergency situations and authorization. It helps in managing several factors that hinder patient analyses for emergencies. A. C. Chang [46] proposed future vision analysis of public health. It is based on diagnosis and therapy analysis for health. It also helps analyze catastrophic systems of diagnosis. It helps in data projection system for deployment and analysis. The method uses a fusion of artificial intelligence, machine learning, and deep learning. It helps mitigate several issues of COVID-19. Suri et al. [47] analyzed a heart and brain analysis based on injury for the patients. In this study, the patient is based on comorbidity disease. For the purpose of this disease, several types of medical images are analyzed to help in the classification system. This classification system is based on severity analysis. The method is based an artificial intelligence technique for detecting pandemic infection. Dorr et al. [48] designed a detection method for pneumonia and its analysis. This analysis is based on chest radiography. The method uses artificial intelligence techniques for chest radiograph analysis with the help of X-ray and its management. The method efficiently analyzes several information based on viral fever for bacterial analysis based on nonpneumonia and pneumonia. The method used is diagnosis and analyzed based on several parameters and helps improve several resource allocation systems. Zhou et al. [49] designed a method of drug repurposing based on an artificial intelligence technique. It is used efficiently to analyze drug information based on requirement and customer satisfaction. The intelligent technique of this system helps in managing network medicine analysis on different cutting-edge systems. It helps in managing several accelerating analyses on repositioning and repurposing drugs. Vaishya et al. [50] designed an application related to COVID-19 based on several diseases and symptoms. The work helps fight against COVID to help the society based on different parameters. It helps reduce workload of different employees in the hospital, making it safe, through analysis of information, for the patients. The work is based on several techniques, such as big data, machine learning, internet of things, and artificial intelligence. The combination of all these methods helps make an intelligent system for COVID-19. Attia et al. [51] designed an analysis system for ECG. The work is based on the dysfunction detection system for COVID. The work is based on a series of case analysis of medical diagnoses and analyses on the ventricles. It checks different types of probabilities of ejection fraction based on artificial intelligence. Rasheed et al. [52] illustrated a survey report on COVID-19. It helps in managing several workers on the front line of health analysis. It consists of several innovations for protecting against the coronavirus. The purpose of this survey is to help illustrate different works based on an artificial intelligence technique. It is based on several types of medical data for pandemic analysis, diagnosis, and prognosis. It is based on the diagnosis of different types of survey report covering multiple disciplines. Mohanty et al. [53] designed an article for an application based on artificial intelligence. The work is based on COVID-19 drug repurposing. It uses several types of strategies based on repositioning of drug to reduce failure or lack during treatment. The model is effective, and its analysis is based on several purposes and trials of different types of effectiveness. Lalmuanawma et al. [54] designed a model based on the fusion of artificial intelligence and machine learning for pandemic analysis of COVID-19. The article is based on a review of papers for the analysis of several hospital information management systems based on the clinical analysis of patient information as well as medicine information. It helps in managing several information on contact tracing, drug analysis, screening, and predictions based on different information. Da Silva et al. [55] designed a method of COVID-19 analysis in the countries of America and Brazil. The purpose of this article is to analyze several information precipitation and temperature based on forecasting model analysis. It helps in managing several accuracies on medical diagnosis based on several factors and management. It is based on public policies for different types of diagnosis. Ke et al. [56] designed a method of fighting with COVID-19 using artificial intelligence. The method uses repurposing of drug for maintaining several analyses. It helps design several types of analysis on anticoronavirus and its management. It uses several learning methods of result analysis on a few drugs. Finally, it helps diagnose several resulting analysis system based on different factors. Ghose et al. [57] designed a method of COVID-19 analysis and diagnosis. It uses artificial intelligence technique for the analysis. It uses several clinical data and information along with several invaluable resources for analyzing. It also helps diagnose several risks to identify different interrupted information. The method uses several statistical analyses for reduction of issues. Vinod and Prabaharan [58] designed a method of COVID-19 analysis and management for fast diagnosis. The work is based on the fusion of artificial intelligence and data science. It uses several datasets for the analysis of several diseases based on different variations and management techniques. It uses decision tree and classifier for analyzing different parameters efficiently and effectively. Shaban et al. [59] designed a patient analysis system used for inference systems and design. It is based on several information that help model several neural network systems. The work is based on a fuzzy inference modeling system that helps manage some issues related to COVID-19. The issue is based on deep learning and fuzzy system modeling. It helps model several network information systems based on strategy management. It helps validate several cross analysis data that help in accuracy modeling. It helps in modeling several detection mechanisms for coronavirus analysis to help model a prevention system. Zahra et al. [60] designed a data-driven system based on intelligence system maintenance and information system. The work is based on the URL system of information and modeling that helps in the analysis and design of several malicious and phishing information systems. The work is completely based on fuzzy logic modeling. It helps deal with several uncertainty management for pandemic analysis. It helps model several control and unprecedented information systems. The work helps model several cybercriminal information analyses to deal with several ransomware information system for impact analysis. In the aforementioned literature, several information is discussed in the form of new innovations and research relating to the critical analysis of the disease based on different variations using different techniques of artificial intelligence [61–63]. Its summarized information is shown in Table 6.1, and its related abbreviations are shown in Table 6.2.

TABLE 6.1
Summarized Information of Literature

Ref.	Year	Purpose	Used Technique
[5]	2021	Digital image analysis for health-care systems	Fusion of artificial intelligence and machine learning techniques
[6]	2021	Patient management	Artificial intelligence
[7]	2021	Student performance prediction	Fusion of artificial intelligence and machine learning techniques
[9]	2021	Cardiology analysis	Artificial intelligence
[10]	2021	Prognostic factor analysis of patients	Artificial intelligence, regression model, ANN
[11]	2021	Medical image processing	Artificial intelligence
[13]	2021	Analysis and prediction of patient information	Artificial intelligence, machine learning, internet of things
[14]	2021	Cardiac sonography analysis	Artificial intelligence
[15]	2021	Analysis the effects of COVID-19	Artificial intelligence
[17]	2021	Analysis of the prognostication of the disease	Artificial intelligence, DNN
[18]	2021	Distance detection by smart camera	Artificial intelligence, CNN
[19]	2021	Tweet fake news detection	Artificial intelligence and classification analysis
[21]	2021	Safety management and drug therapy analysis of COVID-19	Artificial intelligence
[22]	2021	Prediction and severity analysis	Artificial intelligence, random forest, deep learning, GBT
[23]	2021	Dealing battle against COVID-19	Artificial intelligence
[25]	2021	Emergency maintenance system in public heath	Artificial intelligence
[26]	2021	Modeling forecasting in the countries of Brazil and Russia	Artificial intelligence, PSO, ANFIS, MPA
[27]	2021	Clinical care system	Artificial intelligence, MDA
[28]	2021	Syndrome analysis of respiratory distress	Artificial intelligence
[30]	2021	Lung disease analysis based on severity	Artificial intelligence
[31]	2021	Electrochemical immunosensors analysis	Artificial intelligence, IoMT

(Continued)

TABLE 6.1   (Continued)

Ref.	Year	Purpose	Used Technique
[32]	2021	Vaccine distribution and analysis	Artificial intelligence, IoT
[33]	2021	CT image analysis	Artificial intelligence, machine learning, deep learning
[34]	2021	Detection and monitoring information of clinic based on assessment	Artificial intelligence, medical imaging processing
[35]	2021	Enhance interpreter service and multilingual care system	Artificial intelligence
[37]	2021	Discovery of peptide	Artificial intelligence, evolutionary algorithm, machine learning, computational biology
[38]	2021	Pneumonia analysis by using radiological diagnosis	Artificial intelligence, deep learning
[39]	2021	Risk stratification and patient analysis	Artificial intelligence
[40]	2021	Clinical data analysis and routine management	Artificial intelligence, machine learning
[41]	2021	Radiation analysis for pneumonitis diagnosis	Artificial intelligence, deep learning
[42]	2021	Industrial revolution and analysis	Artificial intelligence
[43]	2021	Reverse engineering	Artificial intelligence, big data
[44]	2021	Cancer diagnosis and analysis	Artificial intelligence
[45]	2021	Savior analysis during COVID	Artificial intelligence
[46]	2020	Therapy and diagnosis	Fusion of artificial intelligence, deep learning, and machine learning
[47]	2020	Heart and brain injury analysis of patients	Artificial intelligence and image processing
[48]	2020	Pneumonia analysis and detection	Artificial intelligence
[49]	2020	Drug analysis and repurposing	Artificial intelligence
[50]	2020	Predicting, screening, tracking COVID-19	Artificial intelligence, machine learning, big data, IoT
[51]	2020	Analyze and diagnostic ventricular dysfunction	Artificial intelligence
[52]	2020	Managing worker and helping in efficient decision making	Artificial intelligence
[53]	2020	Repurposing of drug	Artificial intelligence
[54]	2020	Vaccination and drug analysis and prediction	Artificial intelligence, machine learning
[55]	2020	Strategic planning based on diagnosis	Artificial intelligence, RNN, SVR, QRF, KNN
[56]	2020	Analysis of repurposing drugs	Artificial intelligence
[57]	2020	Fight against COVID	Artificial intelligence
[58]	2020	Fast diagnosis and analysis of COVID	Artificial intelligence, decision tree, classifier
[59]	2021	COVID-19 detection	FL
[60]	2021	Detecting COVID-19 along with attack	FL

TABLE 6.2
List of Abbreviations with Description

Abbreviation	Description
ANFIS	Adaptive Neuro-Fuzzy Inference System
ANN	Artificial Neural Network
CNN	Convolution Neural Network
DNN	Deep Neural Network
FL	Fuzzy Logic
GBT	Gradient Boosted Tree
IoMT	Internet of Medical Things
IoT	Internet of Things
KNN	k-Nearest Neighbors
MDA	Multimodal Data Analysis
MPA	Marine Predators Algorithm
PSO	Particle Swarm Optimization
QRF	Quantile Random Forest
RNN	Regression Neural Network
SVR	Support Vector Regression

7.3 ARTIFICIAL INTELLIGENCE TO MITIGATE AND ANALYZE THE DISEASE

In the aforementioned section, several research articles are analyzed and published based on the COVID-19 pandemic for the purpose of formulation. It is based on several information that help analyze and deal with several parameters of the system. This system is based on issues raised by the COVID-19 pandemic. Each of the mentioned issue is based on variations and parameters of COVID-19 that resolve and deal with an artificial intelligence technique or its variants. The main reason for the selection of artificial intelligence with several types of features is to help analyze the problems. Figure 1 shows the relationship between artificial intelligence, machine learning, and deep learning based on learning concepts. It indicates that machine learning is a subset of artificial intelligence and superset of deep learning. So both are subsets of artificial intelligence technique. The working methodologies of these three techniques are based on rules and an agent-based system related to its environment and parameters. It helps recognize and predict information based on several situations for solving problems.

This technique helps in the diagnosis and analysis of several information systems based on its capabilities and agent component system. These agent and intelligence features of artificial intelligence help in monitoring and in the detection system of several clinics, peptide analysis, radiological diagnosis, routing management for any types of disease, radiation analysis, industrial revolution, cancer detection, brain and heart analysis, vaccination analysis, etc., for the purpose of fighting against coronavirus. This fight is based on several variations of

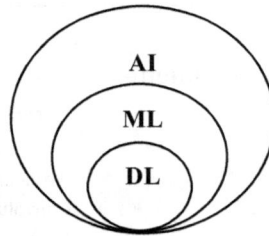

Caption:
AI: Artificial Intelligence
ML: Machine Learning
DL: Deep Learning

FIGURE 7.1 Relationship between artificial intelligence, machine learning, and deep learning.

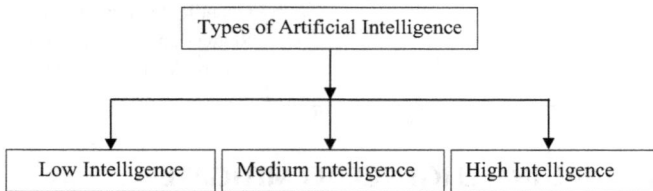

FIGURE 7.2 Categorization of artificial intelligence based on imprecise information.

intelligence for the purpose of reducing uncertainty and helps estimate imprecise information. Because this information and its parameters are not fixed, it changes frequently based on nature or the issues that arise frequently. So based on this critical situation, artificial intelligence is categorized into some types that help analyze information on medical datasets and diseases. This categorization is shown in Figure 7.2. The combination of these three techniques help measure and manage several issues efficiently for problem-solving techniques.

The application of artificial intelligence is based on some modeling and its analysis based on some steps that help in the diagnosis of the problem efficiently. Figure 7.3 shows modeling process of artificial intelligence technique that is used to solve several problems efficiently and effectively for the purpose of innovation. It is used to design new models or the innovation of an existing model based on an intelligence system. It helps solve several issues based on prediction and analysis that is a fusion of multiple agents for solving the stated problem efficiently. It consists of several processes, such as observation, diagnosis, and modeling, applying algorithm, simulating, and predicting. The combination of all these phases helps in the diagnosis and design of an efficient and robust model for handling the situation efficiently.

Figure 7.4 shows the process in mitigating the issues of COVID-19 by agent. This agent is based on an artificial intelligence technique that helps in the modeling

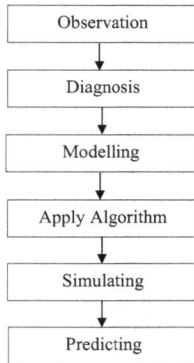

FIGURE 7.3 Modeling process of artificial intelligence.

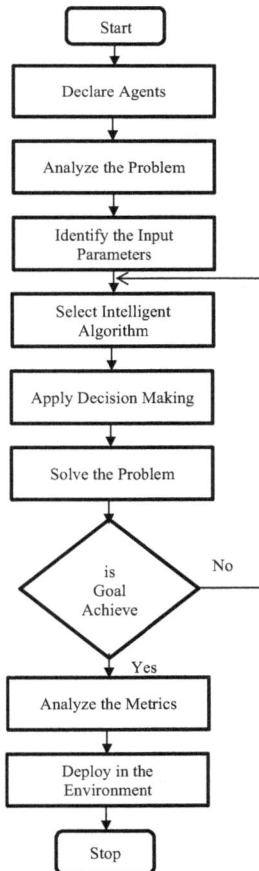

FIGURE 7.4 Process to mitigate the issues of COVID-19 by agent.

of behavior of artificial intelligence. So its working principle is represented by a flowchart system. It is a graphical representation of artificial intelligence techniques for solving any problem. It is based on several agent-based systems that help analyze several problems. The agent helps design the problem based on several input parameters, then it selects algorithms based on the design of the problem and its domain. It helps solve the proposed problem of COVID disease and its different variants efficiently based on decision-making situations. Each disease is based on a goal, and the nature of the goal may be different from its associated metrics. Finally, it helps solve the problem based on several information and data that are available in medical science. Based on metrics, it helps deploy the model into the market to reach out for a solution.

Therefore, the information managed by artificial intelligence is based on several modeling and analysis mentioned in the preceding sections. It efficiently helps deal with each and every information based on its environment. It is associated with several parameters that have an intelligent behavior that helps deal with any type of disease-based data and image analysis and prediction.

7.4   CONCLUSIONS

In this paper, several information are discussed in terms of analysis and design relating to several issues raised by COVID-19 and its variations. Each of these diseases and issues deals with artificial intelligence techniques, such as machine learning, deep neural network, convolution neural network, support vector machine, etc. Each algorithm of artificial intelligence has its own behavior and characteristic for the analysis of issues based on some parameters. It may be single-step analysis or iterative-step analysis. It helps deal with the problem efficiently and also helps in performance analysis based on several metrics. These metrics helps analyze the results that highlight the characteristics of the model.

7.5   REFERENCES

[1] Dones, R. L. E., & Young, M. N. (2020, September). Demand on the of courier services during COVID-19 pandemic in the Philippines. In *2020 7th International Conference on Frontiers of Industrial Engineering (ICFIE)* (pp. 131–134). Piscataway, NJ: IEEE.

[2] Degadwala, S., Vyas, D., & Dave, H. (2021, March). Classification of COVID-19 cases using fine-tune convolution neural network (FT-CNN). In *2021 International Conference on Artificial Intelligence and Smart Systems (ICAIS)* (pp. 609–613). Piscataway, NJ: IEEE.

[3] Das, S. K., Dao, T. P., & Perumal, T. (Eds.). (2021). Nature-inspired computing for smart application design. In *Springer Tracts in Nature-Inspired Computing*, https://doi.org/10.1007/978-981-33-6195-9.

[4] Das, S. K., Das, S. P., Dey, N., & Hassanien, A. E. (Eds.). (2021). *Machine Learning Algorithms for Industrial Applications, Studies in Computational Intelligence*, Springer, https://doi.org/10.1007/978-3-030-50641-4.

[5] Alhasan, M., & Hasaneen, M. (2021). Digital imaging, technologies and artificial intelligence applications during COVID-19 pandemic. *Computerized Medical Imaging and Graphics*, 101933, https://doi.org/10.1016/j.compmedimag.2021.101933.

[6] Lan, L., Sun, W., Xu, D., Yu, M., Xiao, F., Hu, H., & Wang, X. (2021). Artificial intelligence-based approaches for COVID-19 patient management. *Intelligent Medicine*, https://doi.org/10.1016/j.imed.2021.05.005.

[7] Tarik, A., Aissa, H., & Yousef, F. (2021). Artificial intelligence and machine learning to predict student performance during the COVID-19. *Procedia Computer Science*, 184, 835–840.

[8] De, D., Mukherjee, A., Das, S. K., & Dey, N. (Eds.). (2020). Nature inspired computing for wireless sensor networks. In *Springer Tracts in Nature-Inspired Computing*. New York: Springer.

[9] Haleem, A., Javaid, M., Singh, R. P., & Suman, R. (2021). Applications of artificial intelligence (AI) for cardiology during COVID-19 pandemic. *Sustainable Operations and Computers*, 2, 71–78.

[10] Zhang, S., Huang, S., Liu, J., Dong, X., Meng, M., Chen, L., & Chen, D. (2021). Identification and validation of prognostic factors in patients with COVID-19: A retrospective study based on artificial intelligence algorithms. *Journal of Intensive Medicine*, https://doi.org/10.1016/j.jointm.2021.04.001.

[11] Born, J., Beymer, D., Rajan, D., Coy, A., Mukherjee, V. V., Manica, M., & Rosen-Zvi, M. (2021). On the role of artificial intelligence in medical imaging of COVID-19. *Patterns*, https://doi.org/10.1016/j.patter.2021.100269.

[12] Das, S. K., Samanta, S., Dey, N., & Kumar, R. (Eds.). (2020). *Design Frameworks for Wireless Networks, Lecture Notes in Networks and Systems*. Springer, Singapore.

[13] Barnawi, A., Chhikara, P., Tekchandani, R., Kumar, N., & Alzahrani, B. (2021). Artificial intelligence-enabled Internet of Things-based system for COVID-19 screening using aerial thermal imaging. *Future Generation Computer Systems*, https://doi.org/10.1016/j.future.2021.05.019.

[14] Cheema, B. S., Walter, J., Narang, A., & Thomas, J. D. (2021). Artificial intelligence—enabled POCUS in the COVID-19 ICU: A new spin on cardiac ultrasound. *Case Reports*, 3(2), 258–263.

[15] Sucharitha, G., & Chary, D. V. (2021). Predicting the effect of COVID-19 by using artificial intelligence: A case study. *Materials Today: Proceedings*, https://doi.org/10.1016/j.matpr.2021.02.202.

[16] Dey, N. (Ed.). (2021). *Applications of Flower Pollination Algorithm and Its Variants*. Springer, Singapore, ISBN: 978-981-33-6104-1.

[17] Jiao, Z., Choi, J. W., Halsey, K., Tran, T. M. L., Hsieh, B., Wang, D., & Bai, H. X. (2021). Prognostication of patients with COVID-19 using artificial intelligence based on chest x-rays and clinical data: A retrospective study. *The Lancet Digital Health*, 3(5), e286–e294.

[18] Karaman, O., Alhudhaif, A., & Polat, K. (2021). Development of smart camera systems based on artificial intelligence network for social distance detection to fight against COVID-19. *Applied Soft Computing*, 110, 107610, https://doi.org/10.1016/j.asoc.2021.107610.

[19] Madani, Y., Erritali, M., & Bouikhalene, B. (2021). Using artificial intelligence techniques for detecting COVID-19 epidemic fake news in Moroccan tweets. *Results in Physics*, 25, 104266, https://doi.org/10.1016/j.rinp.2021.104266.

[20] Das, S. K. (2021). Smart design and its applications: Challenges and techniques. *Nature-Inspired Computing for Smart Application Design*, 1.

[21] Maille, B., Wilkin, M., Million, M., Rességuier, N., Franceschi, F., Koutbi-Franceschi, L., & Fiorina, L. (2021). Smartwatch electrocardiogram and artificial intelligence for assessing cardiac-rhythm safety of drug therapy in the COVID-19 pandemic. The QT-logs study. *International Journal of Cardiology*, 331, 333–339.

[22] Yaşar, Ş., Çolak, C., & Yoloğlu, S. (2021). Artificial intelligence-based prediction of COVID-19 severity on the results of protein profiling. *Computer Methods and Programs in Biomedicine*, 202, 105996, https://doi.org/10.1016/j.cmpb.2021.105996.

[23] Tayarani-N, M. H. (2020). Applications of artificial intelligence in battling against COVID-19: A literature review. *Chaos, Solitons & Fractals*, 110338, https://doi.org/10.1016/j.chaos.2020.110338.

[24] Mahalle, P. N., Shelar, P. A., Shinde, G. R., & Dey, N. (2021). Introduction to underwater wireless sensor networks. In *The Underwater World for Digital Data Transmission* (pp. 1–21). Springer, Singapore.

[25] Zhu, L., Chen, P., Dong, D., & Wang, Z. (2021). Can artificial intelligence enable the government to respond more effectively to major public health emergencies?-Taking the prevention and control of COVID-19 in China as an example. *Socio-Economic Planning Sciences*, 101029, https://doi.org/10.1016/j.seps.2021.101029.

[26] Al-Qaness, M. A., Saba, A. I., Elsheikh, A. H., Abd Elaziz, M., Ibrahim, R. A., Lu, S., . . . & Ewees, A. A. (2021). Efficient artificial intelligence forecasting models for COVID-19 outbreak in Russia and Brazil. *Process Safety and Environmental Protection*, 149, 399–409, https://doi.org/10.1016/j.psep.2020.11.007.

[27] Adamidi, E. S., Mitsis, K., & Nikita, K. S. (2021). Artificial intelligence in clinical care amidst COVID-19 pandemic: A systematic review. *Computational and Structural Biotechnology Journal*, https://doi.org/10.1016/j.csbj.2021.05.010.

[28] Suri, J. S., Agarwal, S., Gupta, S. K., Puvvula, A., Biswas, M., Saba, L., & Naidu, S. (2021). A narrative review on characterization of acute respiratory distress syndrome in COVID-19-infected lungs using artificial intelligence. *Computers in Biology and Medicine*, 104210, https://doi.org/10.1016/j.compbiomed.2021.104210.

[29] De, D., Mukherjee, A., Das, S. K., & Dey, N. (2020). Wireless sensor network: Applications, challenges, and algorithms. In *Nature Inspired Computing for Wireless Sensor Networks* (pp. 1–18). Springer, Singapore.

[30] Li, M. D., Little, B. P., Alkasab, T. K., Mendoza, D. P., Succi, M. D., Shepard, J. A. O., & Kalpathy-Cramer, J. (2021). Multi-radiologist user study for artificial intelligence-guided grading of COVID-19 lung disease severity on chest radiographs. *Academic Radiology*, 28(4), 572–576.

[31] Yadav, A. K., Verma, D., Kumar, A., Kumar, P., & Solanki, P. R. (2021). The perspectives of biomarkers based electrochemical immunosensors, artificial intelligence and the internet of medical things towards COVID-19 diagnosis and management. *Materials Today Chemistry*, 100443, https://doi.org/10.1016/j.mtchem.2021.100443.

[32] Kumar, R., & Veer, K. (2021). How artificial intelligence and internet of things can aid in the distribution of COVID-19 vaccines. *Diabetes & Metabolic Syndrome*, https://doi.org/10.1016/j.dsx.2021.04.021.

[33] Moezzi, M., Shirbandi, K., Shahvandi, H. K., Arjmand, B., & Rahim, F. (2021). The diagnostic accuracy of Artificial Intelligence-Assisted CT imaging in COVID-19 disease: A systematic review and meta-analysis. *Informatics in Medicine Unlocked*, 100591, https://doi.org/10.1016/j.imu.2021.100591.

[34] Shaikh, F., Andersen, M. B., Sohail, M. R., Mulero, F., Awan, O., Dupont-Roettger, D., & Bisdas, S. (2021). Current landscape of imaging and the potential role for artificial intelligence in the management of COVID-19. *Current Problems in Diagnostic Radiology*, 50(3), 430–435.

[35] Chonde, D. B., Pourvaziri, A., Williams, J., McGowan, J., Moskos, M., Alvarez, C., & Succi, M. D. (2021). RadTranslate: An artificial intelligence—powered intervention for urgent imaging to enhance care equity for patients with limited English proficiency during the COVID-19 pandemic. *Journal of the American College of Radiology*, https://doi.org/10.1016/j.jacr.2021.01.013.

[36] Dey, N., Samanta, S., Chakraborty, S., Das, A., Chaudhuri, S. S., & Suri, J. S. (2014). Firefly algorithm for optimization of scaling factors during embedding of manifold medical information: An application in ophthalmology imaging. *Journal of Medical Imaging and Health Informatics*, 4(3), 384–394.

[37] Kabra, R., & Singh, S. (2021). Evolutionary artificial intelligence based peptide discoveries for effective COVID-19 therapeutics. *Biochimica et Biophysica Acta (BBA)-Molecular Basis of Disease*, 1867(1), 165978, https://doi.org/10.1016/j.bbadis.2020.165978.

[38] Lampejo, T. (2021). Pneumocystis pneumonia: An important consideration when investigating artificial intelligence-based methods in the radiological diagnosis of COVID-19. *Clinical Imaging*, https://doi.org/10.1016/j.clinimag.2021.02.044

[39] Wang, T. K. M., Cremer, P., Chan, N., Piotrowska, H., Woodward, G., & Jaber, W. (2021). Utility of an automated artificial intelligence echocardiography software in risk stratification of hospitalized COVID-19 patients. *Journal of the American College of Cardiology*, 77(18_Supplement_1), 3089–3089.

[40] Soltan, A. A., Kouchaki, S., Zhu, T., Kiyasseh, D., Taylor, T., Hussain, Z. B., & Clifton, D. A. (2021). Rapid triage for COVID-19 using routine clinical data for patients attending hospital: Development and prospective validation of an artificial intelligence screening test. *The Lancet Digital Health*, 3(2), e78–e87.

[41] Ramella, S., Quattrocchi, C. C., Ippolito, E., Giordano, F. M., Greco, C., Mallio, C. A., & Zobel, B. B. (2021). 42P Radiation induced pneumonitis in the era of the COVID-19 pandemic: Artificial intelligence for differential diagnosis. *Journal of Thoracic Oncology*, 16(4), S717, https://doi.org/10.1016/S1556-0864(21)01884-0.

[42] Tiwari, A. K., Abakah, E. J. A., Le, T. L., & Leyva-de la Hiz, D. I. (2021). Markov-switching dependence between artificial intelligence and carbon price: The role of policy uncertainty in the era of the 4th industrial revolution and the effect of COVID-19 pandemic. *Technological Forecasting and Social Change*, 163, 120434, https://doi.org/10.1016/j.techfore.2020.120434.

[43] Haimed, A. M. A., Saba, T., Albasha, A., Rehman, A., & Kolivand, M. (2021). Viral reverse engineering using Artificial Intelligence and big data COVID-19 infection with Long Short-term Memory (LSTM). *Environmental Technology & Innovation*, 22, 101531, https://doi.org/10.1016/j.eti.2021.101531.

[44] Hsieh, J. J. (2021). The 2021 COVID-19 artificial intelligence issue. *Clinical Genitourinary Cancer*, 19(1), 1–2.

[45] Health, T. L. D. (2021). Artificial intelligence for COVID-19: Saviour or saboteur? The Lancet. *Digital Health*, 3(1), e1, https://doi.org/10.1016/S2589-7500(20)30295-8.

[46] Chang, A. C. (2020). Artificial intelligence and COVID-19: Present state and future vision. *Intelligence-Based Medicine*, 3, 100012, https://doi.org/10.1016/j.ibmed.2020.100012.

[47] Suri, J. S., Puvvula, A., Biswas, M., Majhail, M., Saba, L., Faa, G., & Naidu, S. (2020). COVID-19 pathways for brain and heart injury in comorbidity patients: A role of medical imaging and artificial intelligence-based COVID severity classification: A review. *Computers in Biology and Medicine*, 103960, https://doi.org/10.1016/j.compbiomed.2020.103960.

[48] Dorr, F., Chaves, H., Serra, M. M., Ramirez, A., Costa, M. E., Seia, J., & Barmaimon, G. (2020). COVID-19 pneumonia accurately detected on chest radiographs with artificial intelligence. *Intelligence-based Medicine*, 3, 100014, https://doi.org/10.1016/j.ibmed.2020.100014.

[49] Zhou, Y., Wang, F., Tang, J., Nussinov, R., & Cheng, F. (2020). Artificial intelligence in COVID-19 drug repurposing. *The Lancet Digital Health*, https://doi.org/10.1016/S2589-7500(20)30192-8.

[50] Vaishya, R., Javaid, M., Khan, I. H., & Haleem, A. (2020). Artificial Intelligence (AI) applications for COVID-19 pandemic. *Diabetes & Metabolic Syndrome: Clinical Research & Reviews*, 14(4), 337–339.

[51] Attia, Z. I., Kapa, S., Noseworthy, P. A., Lopez-Jimenez, F., & Friedman, P. A. (2020, November). Artificial intelligence ECG to detect left ventricular dysfunction in COVID-19: A case series. In *Mayo Clinic Proceedings* (Vol. 95, No. 11, pp. 2464–2466). Amsterdam, Netherlands: Elsevier.

[52] Rasheed, J., Jamil, A., Hameed, A. A., Aftab, U., Aftab, J., Shah, S. A., & Draheim, D. (2020). A survey on artificial intelligence approaches in supporting frontline workers and decision makers for COVID-19 pandemic. *Chaos, Solitons & Fractals*, 110337, https://doi.org/10.1016/j.chaos.2020.110337.

[53] Mohanty, S., Rashid, M. H. A., Mridul, M., Mohanty, C., & Swayamsiddha, S. (2020). Application of artificial intelligence in COVID-19 drug repurposing. *Diabetes & Metabolic Syndrome: Clinical Research & Reviews*, https://doi.org/10.1016/j.dsx.2020.06.068.

[54] Lalmuanawma, S., Hussain, J., & Chhakchhuak, L. (2020). Applications of machine learning and artificial intelligence for COVID-19 (SARS-CoV-2) pandemic: A review. *Chaos, Solitons & Fractals*, 139, 110059, https://doi.org/10.1016/j.chaos.2020.110059.

[55] da Silva, R. G., Ribeiro, M. H. D. M., Mariani, V. C., & dos Santos Coelho, L. (2020). Forecasting Brazilian and American COVID-19 cases based on artificial intelligence coupled with climatic exogenous variables. *Chaos, Solitons & Fractals*, 139, 110027, https://doi.org/10.1016/j.chaos.2020.110027.

[56] Ke, Y. Y., Peng, T. T., Yeh, T. K., Huang, W. Z., Chang, S. E., Wu, S. H., & Chen, C. T. (2020). Artificial intelligence approach fighting COVID-19 with repurposing drugs. *Biomedical Journal*, 43(4), 355–362.

[57] Ghose, A., Roy, S., Vasdev, N., Olsburgh, J., & Dasgupta, P. (2020). The emerging role of Artificial Intelligence in the fight against COVID-19. *European Urology*, 78(6), 775, https://doi.org/10.1016/j.eururo.2020.09.031.

[58] Vinod, D. N., & Prabaharan, S. R. S. (2020). Data science and the role of Artificial Intelligence in achieving the fast diagnosis of COVID-19. *Chaos, Solitons & Fractals*, 140, 110182, https://doi.org/10.1016/j.chaos.2020.110182.

[59] Shaban, W. M., Rabie, A. H., Saleh, A. I., & Abo-Elsoud, M. A. (2021). Detecting COVID-19 patients based on fuzzy inference engine and Deep Neural Network. *Applied Soft Computing*, 99, 106906, https://doi.org/10.1016/j.asoc.2020.106906.

[60] Zahra, S. R., Chishti, M. A., Baba, A. I., & Wu, F. (2021). Detecting COVID-19 chaos driven phishing/malicious URL attacks by a fuzzy logic and data mining based intelligence system. *Egyptian Informatics Journal*, https://doi.org/10.1016/j.eij.2021.12.003.

[61] Ahmed, H. I., Nasr, A. A., Abdel-Mageid, S. M., & Aslan, H. K. (2021). DADEM: Distributed attack detection model based on big data analytics for the enhancement of the security of internet of things (IoT). *International Journal of Ambient Computing and Intelligence (IJACI)*, 12(1), 114–139.

[62] Sholla, S., Mir, R. N., & Chishti, M. A. (2021). A fuzzy logic-based method for incorporating ethics in the internet of things. *International Journal of Ambient Computing and Intelligence (IJACI)*, 12(3), 98–122.

[63] Balusa, B. C., & Gorai, A. K. (2021). Development of fuzzy pattern recognition model for underground metal mining method selection. *International Journal of Ambient Computing and Intelligence (IJACI)*, 12(4), 64–78.

8 Data Analysis and Prediction for WSN Based on Linear and Quadratic Optimization Techniques

Manoj Kumar Mandal, Arun Prasad Burnwal, B. K. Mahatha, and Abhishek Kumar

CONTENTS

8.1 Introduction 137
8.2 Related Works.. 138
8.3 Preliminaries 142
    8.3.1 Mathematical Modeling... 142
    8.3.2 Linear Programming 142
    8.3.3 Quadratic Programming.. 142
    8.3.4 Fuzzy Logic 142
8.4 Proposed Method...143
8.5 Simulation and Performance Analysis 148
8.6 Conclusions... 152
8.7 References.. 152

8.1 INTRODUCTION

The wireless sensor network (WSN) is made up of sensor nodes and base stations (BS). The sensor nodes' role is to detect environmental information and transmit it to the BS. This information is processed by the BS, which predicts the user inquiry and responds appropriately [1-2]. The WSN's key parameter is energy, which efficiently reflects on other parameters. Because each sensor node has a limited battery capacity, this low energy capacity is insufficient for each user's activity. Some work is based on the different challenges of management systems for security issues of the information system within the context of the wireless network. Some of the operation is based on coverage optimization and metaheuristic optimization that helps model several applications [3-4]. During the operation, sometimes nodes fail to transmit the data packet, or the path between nodes fails due to a lack of required energy. This issue affects other network parameters efficiently, like increase in some network metrics, such as packet delivery ratio, throughput, and goodput, and decreases in some network metrics, such as packet loss, overhead, and end-to-end delay. The combining variation of both types of network parameters degrades the network lifetime and affects the overall operation of the network.

The proposed method basically discusses one type of network parameter needed to be minimized for packet loss, overhead, and end-to-end delay. The combination of the three minimizations helps enhance the network lifetime. In this paper, the main mathematical modeling is a fusion of linear programming and quadratic programming. The combination of both mathematical modeling helps analyze the network data and predict the results for verification and to arrive at a conclusion.

The remainder of the paper discusses several works based on existing works. The next work is based on preliminaries of the works for the purpose of the main work. The next work discusses the main proposed method. And the last section describes the final conclusion of the paper.

8.2   RELATED WORKS

Several works have been proposed for the wireless ad hoc network as well as the wireless sensor network. Each work is based on a real-life purpose and its management. It helps deal with several operations of the sensor network. Some of the works mentioned in this section include a proposed method by S. K. Das [5] for an application design system and its management. It helps in several application management dealing with several challenges and issues. It helps model several information systems based on smart applications. It helps model several security management systems based on emergency management and applications. It helps give a new guideline that helps model several issues in terms of solution. This solution helps in several services based on real-life application management. It helps in several monitoring and application management systems based on emergency and security modeling systems on services. N. Dey et al. [6] designed a method of big data analysis and modeling. It helps model several next-generation information systems. It helps in modeling several intelligence applications based on different types of services. The works of this book are based on the fusion of the internet of things, cloud computing, and several intelligence systems. It helps model new techniques and services based on new and smart applications. It helps model big data analytics that help model some internet of things information systems. Keerthika and Shanmugapriya [7] designed a method that is a combination of passive and active attack for the purpose of illustration. This illustration is based on some countermeasure system that helps in vulnerabilities system. It helps model the application based on an environment analysis dealing with a protection system based on commercial analysis. The communication of the system is based on the deployment of some challenges along with issues. It helps in defensive analysis based on vulnerability based on some factors of information. Wan and Chen [8] designed a strategy for energy analysis and mechanism for harvesting analysis. The work is based on the WSN purpose of modeling. It helps model several cooperative analyses for node analysis. It defines some probabilities based on relay node detection. The main purpose of this analysis is to solve network performance based on certain factors. It helps model the application and save actual energy. It uses mathematical modeling for analyzing data and its parameters. It helps enhance energy based on a solar energy system and its cooperation. N. Dey [9] proposed several methods and operations based on firefly algorithms that help design several types of services. It helps model several embedded and medical analysis system. The work is based on application and management that helps model several imaging systems. The works of this book are based on some application management systems that help in managing several metaheuristic systems. It helps in managing several decomposition analyses based on image analysis. This analysis is based on the fusion of several embedded information systems. It helps in managing several electronic patient record information systems based on nature-inspired systems and modeling. Das et al. [10] designed a book for the purpose of smart application. The work of this book is based on smart application with the fusion of smart computing. It helps model several applications based on some services and management. It helps model several nature-inspired applications based on computing applications. It helps model and give new insight into the subarea of network modeling, data analysis and prediction, network lifetime management, resource and energy management, etc. It helps model several information systems for managing dynamic applications and planning and services. Misra et al. [11] designed an implementation method based on the fusion of FPGA and NLOS. The work is based on distance analysis and its estimation system. It helps model several applications that help in elderly modeling. The work is designed for indoor systems that help in WSN. It helps in location analysis based on the ZigBee network. The work uses programmable gate array system and its modeling. This modeling uses artificial neural network to estimate different errors and improve network lifetime. It uses hybridization method for modeling several complexities based on suitable analysis. Wang and Hu [12] designed a hole-detection method for handling several issues based on WSN. The network is based on clustering methods and algorithms that use some gap-coverage analysis. It helps analyze multihop management systems for rational deployment. It helps distance parameters system and vulnerability detection, which helps in coverage and its parameters modeling. It overcomes the limitation of several determination systems for edge node modeling. It helps determine random walk connection and its management. De et al. [13] proposed an illustration for the purpose of challenges and application management services. This service is based on the application of a wireless sensor network system and its variations. It helps model several challenges and application management. It helps deal with several algorithms of the wireless network based on variation and its analysis. It helps guide several working principles and information systems based on service management. It helps deal with algorithm analysis that helps in managing several applications of the system. Temene et al. [14] illustrated a survey based on mobility analysis and prediction for WSN. The work is based on IoT and WSN both for detailed illustration. It helps model several mobile nodes. There are several mobile nodes that play different roles, such as the sink node, mobile node, source node, etc. The combination of all nodes helps model several congestions and its related mitigations. It helps in the predecessor analysis of IoT, which helps in several directions. The work helps model several evaluations based on different algorithms. Yousefpoor et al. [15] designed a secure method for WSN as a review paper that helps model several issues in the network. The work is based on a data aggregation method that helps reduce attack in the system. It helps in countermeasures for several issues within the context of attack measurement. This review is also based on industrial internet of things and its modeling. It helps in managing several issues within the context of solution measurement. It helps save energy and increase security of the system based on authentication system. Das et al. [16] designed a book of architectural solution system based on wireless network. This service is not only based on network but also on several systems and information based on the architecture of the network. It helps in modifications based on the architecture of the system. It helps model several issues of the network. Several issues are used and deal with the system. Some of the issues mentioned include energy efficiency system, network lifetime system, resource management, data aggregation system, etc. It helps model several solutions and security management of wireless network. Zhang and Mao [17] designed a multifactor system for authentic purposes. The work is based on a protocol system that helps model the application. It helps model several recognitions to exercise the physical system. It is based on the ZigBee network, which helps model several scope identifications. It helps in security analysis and recognition of several applications based on component analysis and its management. It helps model several information based on radio frequency analysis. It helps design the system based on security analysis for connection management of the network. Huanan et al. [18] designed a security-based application system for the wireless sensor network. It helps model several systems for handling several intrusion and detection systems. It helps model several systems based on the foundation of the network. It helps in the study of the system for reducing several threats. It helps model some analysis and emphasis in communication and modeling. It helps handle several security systems for designing analysis and its modeling. De et al. [19] designed a book for the purpose of wireless sensor network. It helps model several applications based on services and management. It helps deal with several information systems and management for key area management. The work is based on nature-inspired application and computing that helps model several issues. It helps implement several applications and computation information systems. Information of this book is distributed in the form of bio- and nature-inspired systems. It helps model and design several applications for single-objective and multiobjective optimization systems. Anand et al. [20] designed a framework system for managing several applications based on multicast service. It helps in managing several protocol systems based on service management. The work is based on dependability analysis and productivity management, which helps in single transmission. The work helps model several retreating systems to manage transparency. It helps deal with several confident and legitimate analyses to manage node along with network information. It helps in managing several intrusions and increases network lifetime. Tahir et al. [21] designed a clustering system to manage several communication systems based on peer-to-peer network and management. The work helps deal with several overlay management on application systems. The clustering system of the network is based on multiple dimensional analysis. It helps in managing several linkages and its analysis that help in lookup management. It helps decrease several complexities, such as overhead, computation system, error, etc. Finally, it helps model several environmental analyses based on path management. Das et al. [22] designed a system and modeling for the purpose of wireless network and application. It helps model several information based on service management and application modeling. The work is based on several information and management systems, such as energy resource management and modeling. It helps deal with several security and privacy management systems that help in its design and enhancement. The work is based on troubleshooting and automation system for network lifetime management and its enhancement. It gives several protocols for the purpose of design perspective and modeling. Sharma and Kim [23] designed a method for multipath management that helps model several information. It helps model several applications based on routing information. This information helps model the network. This network is based on application management of mobile ad hoc network. The work uses a bioinspired technique that helps in managing some applications based on services. It helps model several constraints, such as low memory management, bandwidth, battery life, etc. The combination of all information helps model several applications along with services to adjust the model of the network. Lee et al. [24] designed a technique that models several sharing information systems. The work is based on a military application that uses mobile ad hoc network to helps in managing some applications. The proposed work is termed as cooperative phase with steering system based on relay management. It consists of several relay and destination node management based on services. The network is also attached with a cognitive network model to increase the probability of routing. It helps select relay based on several source nodes to increase the performance of the model. Das et al. [25] designed an application management illustration for the purpose of wireless network system and services. It helps model several applications based on different security and challenge system management. It helps in issue management based on several parts of the communication system. It helps adopt several solutions based on some higher analysis and management. The work is based on complexity management, which helps model several solutions within the context of management based on different variations of the wireless network. It is used for several types of optimization, security, and learning systems [26–28]. Singh et al. [29] proposed an adaptive method for an energy-aware system that is used for communication purposes. This method is used for a physical communication system that is based on a delay-tolerant system. It helps model applications based on a cyber-physical system that is used for wireless sensor network. It helps model applications based on delay-tolerant and prediction system. The application is based on mobile applications for operating several operations. The work is simulated based on the network environment for predicting network lifetime. Singh and Pamula [30] designed a method of intelligent communication and route modeling. The work is based on a vehicular communication system based on strategy management. It helps, based on strategy behavior modeling and analysis, enhance network lifetime. The application is deployed for several purposes based on protocol management. It helps track analysis of novel behavior based on delay-tolerant management. It helps utilize and analyze the system based on vehicular communication. It helps outperform the result based on several variations.

8.3 PRELIMINARIES

Some preliminary steps are discussed in this part to assist you in understanding the basic, important aspects of the suggested strategy.

8.3.1 MATHEMATICAL MODELING

Mathematical modeling is a process to convert the human thinking into terms of mathematics which help code or decode the problem and its related solution efficiently. There are several techniques used in mathematical modeling, such as traditional mathematics and some numerical optimization, including linear programming and its several extended variations. In this paper, two optimization models are used, namely, linear programming and quadratic programming.

8.3.2 LINEAR PROGRAMMING

Linear programming is used in modeling the goal of the problem with the help of related constraints. It includes several elements, such as objective function, constraints, and decision variable. The combined nature of the linear programming is linear, and the degree of the decision variable is one with respect to constraints.

8.3.3 QUADRATIC PROGRAMMING

Quadratic programming is an extended form of linear programming that contains objective functions, constraints, and related decision variables. In this mathematical model, the degree of the decision variable is two, so in this mathematical model, output varies from linear programming. The main objective value increases.

8.3.4 FUZZY LOGIC

Fuzzy logic is a multivalue logic that is used in soft computing to deal with partial truth and partial falsehood using the degree of membership value. Fuzzy logic includes a variable known as a linguistic variable, which deals with imprecise information. The objective of these variables is to lessen the problem's uncertainty and get to the goal.

8.4 PROPOSED METHOD

The main proposed technique is demonstrated in this section using fundamental foundations and mathematical modeling. The purpose of this method is to analyze and predict the data of the WSN based on the fusion of linear and quadratic programming. In this proposed method, there are three objectives considered, as shown in Table 8.1, where packet loss is considered as 100 unit, overhead considered as 200 unit, and end-to-end delay considered as 300 unit. The membership function of these three objectives is illustrated in Tables 8.2 to 8.4 for handling uncertainty and reducing imprecise information.

TABLE 8.1
List of Objectives for Minimization

Sl.no.	Objective
1	Packet loss
2	Overhead
3	End-to-end delay

TABLE 8.2
Membership Function of Packet Loss

Linguistic Variable	Range
Low	(0–40)
Medium	(30–80)
High	(60–100)

TABLE 8.3
Membership Function of Overhead

Linguistic Variable	Range
Low	(0–80)
Medium	(60–170)
High	(140–200)

TABLE 8.4
Membership Function of End-to-End Delay

Linguistic Variable	Range
Low	(0–140)
Medium	(120–200)
High	(190–300)

The mathematical model of the linear programming is shown in equations 1 to 5, and for quadratic programming, it is shown in equations 6 and 10. In these two mathematical models, p_i, o_i, and e_i indicate the three network metrics, namely, packet loss, overhead, and end-to-end delay, for minimization purpose, where i varies three times 1 for "low," 2 for "medium," and 3 for "high" linguistic variables. And x_1, x_2, and x_3 are three decision variables for controlling the three objectives. The two mathematical models are evaluated in 5 times under different sensor nodes, such as 1,000 nodes, 2,000 nodes, 3,000 nodes, 4,000 nodes, and 5,000 nodes. The summarized datasets for the linear model are shown in Tables 8.5 to 8.7 using "low," "medium," and "high" linguistic variables. And Tables 8.8 to 8.10 show datasets of the quadratic model under "low," "medium," and "high" linguistic variables.

TABLE 8.5
Dataset for Linear Model Under 1,000 to 5,000 Sensor Nodes for "Low" Linguistic Variable

Metrics	Sensor Nodes				
	1,000	2,000	3,000	4,000	5,000
x_1	10.00000	0.00000	0.00000	0.000000	0.000000
x_2	15.00000	22.22222	38.18182	60.71429	42.30769
x_3	0.000000	7.407407	32.72727	14.28571	46.15385
Objective Value	25.00000	29.62963	70.90909	75.00000	88.46154

TABLE 8.6
Dataset for Linear Model Under 1,000 to 5,000 Sensor Nodes for "Medium" Linguistic Variable

Metrics	Sensor Nodes				
	1,000	2,000	3,000	4,000	5,000
x_1	0.000000	0.000000	0.000000	0.000000	0.000000
x_2	0.000000	0.000000	0.000000	0.000000	0.000000
x_3	8.333333	15.38462	21.42857	32.00000	40.00000
Objective Value	8.333333	15.38462	21.42857	32.00000	40.00000

TABLE 8.7

Dataset for Linear Model Under 1,000 to 5,000 Sensor Nodes for "High" Linguistic Variable

Metrics	Sensor Nodes				
	1,000	2,000	3,000	4,000	5,000
x_1	0.000000	0.000000	0.000000	0.000000	0.000000
x_2	0.000000	0.000000	0.000000	0.000000	0.000000
x_3	5.263158	10.25641	15.78947	20.51282	25.00000
Objective Value	5.263158	10.25641	15.78947	20.51282	25.00000

TABLE 8.8

Dataset for Quadratic Model Under 1,000 to 5,000 Sensor Nodes for "Low" Linguistic Variable

Metrics	Sensor Nodes				
	1,000	2,000	3,000	4,000	5,000
x_1	9.756559	7.317093	18.18169	20.61849	14.22382
x_2	9.756171	19.51217	30.90914	49.48407	35.79756
x_3	7.316386	7.317121	29.09093	20.61979	46.81033
Objective Value	243.9024	487.8049	2132.231	3298.969	3674.989

TABLE 8.9

Dataset for Quadratic Model Under 1,000 to 5,000 Sensor Nodes for "Medium" Linguistic Variable

Metrics	Sensor Nodes				
	1,000	2,000	3,000	4,000	5,000
x_1	1.587122	3.418832	5.263100	9.617406	6.278140
x_2	3.174603	5.982896	8.421008	11.36606	13.45298
x_3	6.349251	11.11111	14.73689	21.85799	26.00891
Objective Value	52.91005	170.9402	315.7895	699.4536	896.8610

TABLE 8.10

Dataset for Quadratic Model Under 1,000 to 5,000 Sensor Nodes for "High" Linguistic Variable

Metrics	Sensor Nodes				
	1,000	2,000	3,000	4,000	5,000
x_1	1.0118607	2.139864	3.422050	4.691267	5.089330
x_2	2.300875	4.585389	7.072392	9.106559	11.34774
x_3	3.204048	5.961032	8.669081	10.76239	13.75506
Objective Value	16.83341	61.13871	136.8821	220.658	343.8742

Minimize: $\quad Z_1 = x_1 + x_2 + x_3;$

Subject to constraints:
$$p_1x_1 + o_1x_2 + e_1x_3 \geq 1,000;$$
$$p_2x_1 + o_2x_2 + e_2x_3 \geq 1,000;$$
(8.1)
$$p_3x_1 + o_3x_2 + e_3x_3 \geq 1,000;$$
$p_i, o_i,$ and e_i for Low, Medium, and High
$$i = 1, 2, 3$$

Minimize: $\quad Z_2 = x_1 + x_2 + x_3;$

Subject to constraints:
$$p_1x_1 + o_1x_2 + e_1x_3 \geq 2,000;$$
$$p_2x_1 + o_2x_2 + e_2x_3 \geq 2,000;$$
(8.2)
$$p_3x_1 + o_3x_2 + e_3x_3 \geq 2,000;$$
$p_i, o_i,$ and e_i for Low, Medium, and High
$$i = 1, 2, 3$$

Minimize: $\quad Z_3 = x_1 + x_2 + x_3;$

Subject to constraints:
$$p_1x_1 + o_1x_2 + e_1x_3 \geq 3,000;$$
$$p_2x_1 + o_2x_2 + e_2x_3 \geq 3,000;$$
(8.3)
$$p_3x_1 + o_3x_2 + e_3x_3 \geq 3,000;$$
$p_i, o_i,$ and e_i for Low, Medium, and High
$$i = 1, 2, 3$$

Minimize: $\quad Z_4 = x_1 + x_2 + x_3;$

Subject to constraints:
$$p_1x_1 + o_1x_2 + e_1x_3 \geq 4,000;$$
$$p_2x_1 + o_2x_2 + e_2x_3 \geq 4,000;$$
(8.4)
$$p_3x_1 + o_3x_2 + e_3x_3 \geq 4,000;$$
$p_i, o_i,$ and e_i for Low, Medium, and High
$$i = 1, 2, 3$$

Minimize: $\quad Z_5 = x_1 + x_2 + x_3;$

Subject to constraints:
$$p_1x_1 + o_1x_2 + e_1x_3 \geq 5,000;$$
$$p_2x_1 + o_2x_2 + e_2x_3 \geq 5,000;$$
(8.5)

$$p_3x_1 + o_3x_2 + e_3x_3 \geq 5,000;$$
p_i, o_i, and e_i for Low, Medium, and High
$$i = 1, 2, 3$$

Minimize: $\quad Z_6 = (x_1)^2 + (x_2)^2 + (x_3)^2;$

Subject to constraints: $\quad p_1x_1 + o_1x_2 + e_1x_3 \geq 1,000;$

$$p_2x_1 + o_2x_2 + e_2x_3 \geq 1,000; \qquad (8.6)$$

$$p_3x_1 + o_3x_2 + e_3x_3 \geq 1,000;$$

p_i, o_i, and e_i for Low, Medium, and High
$$i = 1, 2, 3$$

Minimize: $\quad Z_7 = (x_1)^2 + (x_2)^2 + (x_3)^2;$

Subject to constraints: $\quad p_1x_1 + o_1x_2 + e_1x_3 \geq 2,000;$

$$p_2x_1 + o_2x_2 + e_2x_3 \geq 2,000; \qquad (8.7)$$

$$p_3x_1 + o_3x_2 + e_3x_3 \geq 2,000;$$

p_i, o_i, and e_i for Low, Medium, and High
$$i = 1, 2, 3$$

Minimize: $\quad Z_8 = (x_1)^2 + (x_2)^2 + (x_3)^2;$

Subject to constraints: $\quad p_1x_1 + o_1x_2 + e_1x_3 \geq 3,000;$

$$p_2x_1 + o_2x_2 + e_2x_3 \geq 3,000; \qquad (8.8)$$

$$p_3x_1 + o_3x_2 + e_3x_3 \geq 3,000;$$

p_i, o_i, and e_i for Low, Medium, and High
$$i = 1, 2, 3$$

Minimize: $\quad Z_9 = (x_1)^2 + (x_2)^2 + (x_3)^2;$

Subject to constraints: $\quad p_1x_1 + o_1x_2 + e_1x_3 \geq 4,000;$

$$p_2x_1 + o_2x_2 + e_2x_3 \geq 4,000; \qquad (8.9)$$

$$p_3x_1 + o_3x_2 + e_3x_3 \geq 4,000;$$

p_i, o_i, and e_i for Low, Medium, and High
$$i = 1, 2, 3$$

Minimize: $\quad Z_{10} = (x_1)^2 + (x_2)^2 + (x_3)^2;$

Subject to constraints: $\quad p_1x_1 + o_1x_2 + e_1x_3 \geq 5,000;$

$$p_2x_1 + o_2x_2 + e_2x_3 \geq 5,000; \qquad (8.10)$$

$$p_3x_1 + o_3x_2 + e_3x_3 \geq 5,000;$$

p_i, o_i, and e_i for Low, Medium, and High
$$i = 1, 2, 3$$

From the mathematical modeling of equations 8.1 to 8.10, Tables 8.5 to 8.10 are generated. So it is concluded that in the quadratic model, the objective value increases more than the linear model based on the nature of the linguistic variables. But it is observed that in each mathematical model, such as the linear and quadratic models, as the behavior of linguistic variables increases, the value of the objective function decreases.

8.5 SIMULATION AND PERFORMANCE ANALYSIS

The specifics of simulation and analysis are illustrated in this section. The suggested strategy is tested using the LINGO optimization simulator, which is based on a combination of linear and nonlinear formulations. Table 8.11 displays the fundamental simulation settings. Windows 11 is utilized in this simulation, along with MS Office 2016 and an optimization software. The suggested technique employs ten mathematical models which are both linear and nonlinear in character. The total linear model used is 5, and the total nonlinear model used is 5. The proposed method is evaluated in five iterations based on the number of nodes as 1,000, 2,000, 3,000, 4,000, and 5,000. So here the minimum sensor node is 1,000 and the maximum sensor node is 5,000. The total objective function used is 10 for the combination of linear and quadratic programming. In linear programming, five objective functions are used with 3×5 constraints, and in quadratic programming, five objectives are also used with 3×5 constraints. The total linguistic variables used is 3, listed as "low," "medium," and "high,"

Figures 8.1 to 8.3 show the illustration of the linear model based on wireless sensor nodes 1,000, 2,000, 3,000, 4,000, and 5,000 in linear environment. In this model, network lifetime is shown as objective value for each linguistic variable, namely, "low," "medium," and "high," when the number of nodes is maximum, i.e., 5,000. It is observed that as the number of nodes increases, then network lifetime also increases. Figure 8.1 shows network lifetime at maximum point under "low" linguistic variable, which is 88.46154. Figure 8.2 shows network lifetime at maximum point under "medium" linguistic variable, which is 40.00000. Figure 8.3 shows network lifetime at maximum point under "high" linguistic variable, which is 25.00000.

Figures 8.4 to 8.6 show an illustration of the quadratic model based on wireless sensor nodes 1,000, 2,000, 3,000, 4,000, and 5,000 in a quadratic environment.

TABLE 8.11
Details of Simulation Parameters

Parameter	Description
Total optimization models	10
Linear model	5
Nonlinear model	5
Windows	Windows 11
Optimization software	LINGO
MS Office	2016
Minimum number of nodes	1,000
Maximum number of nodes	5,000
Total objective functions	10
Nature of the objectives	Linear and nonlinear
Total constraints	30
Constraints for linear model	3×5
Constraints for nonlinear model	3×5
Total linguistic variables	3
Name of the linguistic variables	Low, medium, and high

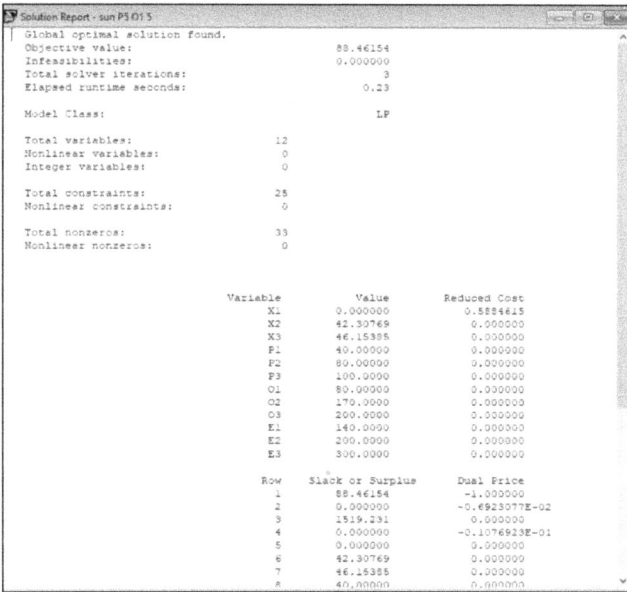

FIGURE 8.1 Maximum network lifetime based on 5,000 nodes under "low" fuzzy variable in a linear environment.

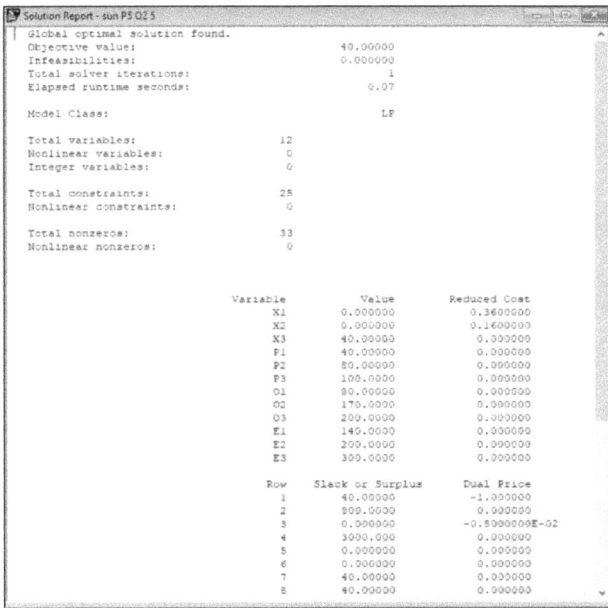

FIGURE 8.2 Maximum network lifetime based on 5,000 nodes under "medium" fuzzy variable in a linear environment.

```
Solution Report - sun PS O3 5
Global optimal solution found.
  Objective value:                      25.00000
  Infeasibilities:                       0.000000
  Total solver iterations:                      1
  Elapsed runtime seconds:                   0.07

  Model Class:                                 LP

  Total variables:             12
  Nonlinear variables:          0
  Integer variables:            0

  Total constraints:           25
  Nonlinear constraints:        0

  Total nonzeros:              33
  Nonlinear nonzeros:           0


              Variable          Value      Reduced Cost
                    X1       0.000000         0.6300000
                    X2       0.000000         0.1750000
                    X3       25.00000         0.000000
                    P1       40.00000         0.000000
                    P2       80.00000         0.000000
                    P3       100.0000         0.000000
                    O1       80.00000         0.000000
                    O2       170.0000         0.000000
                    O3       200.0000         0.000000
                    E1       140.0000         0.000000
                    E2       200.0000         0.000000
                    E3       300.0000         0.000000

                Row #   Slack or Surplus     Dual Price
                    1       25.00000        -1.000000
                    2        0.000000        -0.5000000E-02
                    3        1625.000        0.000000
                    4        2000.000        0.000000
                    5        0.000000        0.000000
                    6        0.000000        0.000000
                    7        25.00000        0.000000
                    8        40.00000        0.000000
```

FIGURE 8.3 Maximum network lifetime based on 5,000 nodes under "high" fuzzy variable in a linear environment.

```
Solution Report - sun PS Q1 5
Global optimal solution found.
  Objective value:                      3674.989
  Infeasibilities:                     0.8497405E-08
  Total solver iterations:                     10
  Elapsed runtime seconds:                   0.07
  Model is convex quadratic

  Model Class:                                 QP

  Total variables:             12
  Nonlinear variables:          3
  Integer variables:            0

  Total constraints:           25
  Nonlinear constraints:        1

  Total nonzeros:              33
  Nonlinear nonzeros:           3


              Variable          Value      Reduced Cost
                    X1       14.22382        -0.7304683E-03
                    X2       35.79756        -0.1599222E-02
                    X3       46.81033         0.1444777E-02
                    P1       40.00000         0.000000
                    P2       80.00000         0.000000
                    P3       100.0000         0.000000
                    O1       80.00000         0.000000
                    O2       170.0000         0.000000
                    O3       200.0000         0.000000
                    E1       140.0000         0.000000
                    E2       200.0000         0.000000
                    E3       300.0000         0.000000

                Row    Slack or Surplus     Dual Price
                    1       3674.989        -1.000000
                    2      -0.8497405E-08   -0.7667157
                    3       1505.280        -0.6334758E-08
                    4       0.1210837E-05   -0.7032800
                    5       14.22382         0.000000
                    6       35.79756         0.000000
                    7       46.81033         0.000000
                    8       40.00000         0.000000
```

FIGURE 8.4 Maximum network lifetime based on 5,000 nodes under "low" fuzzy variable in a quadratic environment.

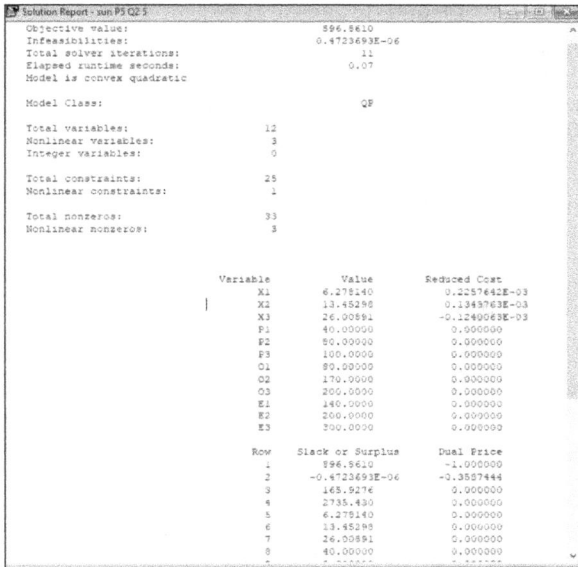

FIGURE 8.5   Maximum network lifetime based on 5,000 nodes under "medium" fuzzy variable in a quadratic environment.

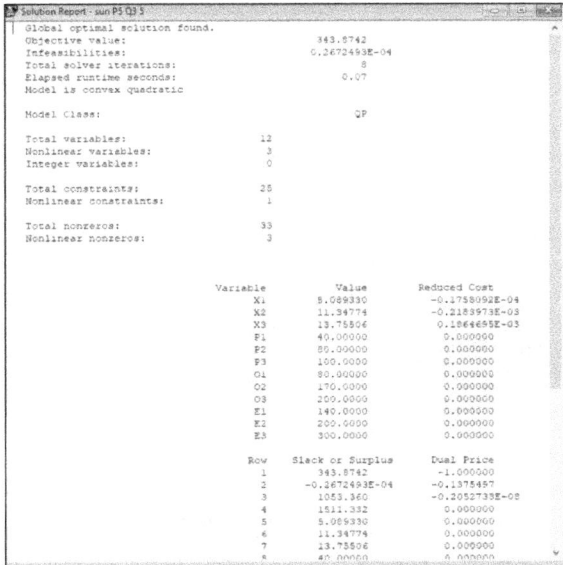

FIGURE 8.6   Maximum network lifetime based on 5,000 nodes under "high" fuzzy variable in a quadratic environment.

In this model, network lifetime is shown as objective value for each linguistic variable, namely, "low," "medium," and "high," when the number of nodes is maximum, i.e., 5,000. It is observed that as the number of nodes increases, then network lifetime also increases. Figure 8.4 shows network lifetime at maximum point under "low" linguistic variable, which is 3674.989. Figure 8.5 shows network lifetime at maximum point under "medium" linguistic variable, which is 896.8610. Figure 8.6 shows network lifetime at maximum point under "high" linguistic variable, which is 343.8742.

8.6 CONCLUSIONS

In this research, an effective approach for modeling and prediction is provided. The purpose of this technique is to enhance network lifetime of WSN by optimizing three metrics, namely, packet loss, overhead, and end-to-end delay. These three metrics are minimized within the context of the fusion of mathematical modeling and fuzzy logic. The proposed method is analyzed in three linguistic variables, namely, "low," "medium," and "high," with different scenarios based on variations of sensor nodes 1,000 to 5,000. The network lifetime of WSN decreases as the nature of the linguistic variables increases. And network lifetime of WSN increases as the number of nodes increases. The combination of both strategies helps in handling uncertainties of the network efficiently. Finally, it is observed that the combination of linear programming, quadratic programming, and fuzzy logic efficiently helps in data analysis and prediction of the network.

8.7 REFERENCES

[1] Keerthika, M., & Shanmugapriya, D. (2021). Wireless sensor networks: Active and passive attacks-vulnerabilities and countermeasures. *Global Transitions Proceedings*, 2(2), 362–367.
[2] Wan, J., & Chen, J. (2022). AHP based relay selection strategy for energy harvesting wireless sensor networks. *Future Generation Computer Systems*, 128, 36–44.
[3] Stevovic, I., Mirjanic, D., & Petrovic, N. (2021). Integration of solar energy by nature-inspired optimization in the context of circular economy. *Energy*, 235, 121297, https://doi.org/10.1016/j.energy.2021.121297.
[4] Li, H., Huang, Z., Liu, X., Zeng, C., & Zou, P. (2020). Multi-fidelity meta-optimization for nature inspired optimization algorithms. *Applied Soft Computing*, 96, 106619, https://doi.org/10.1016/j.asoc.2020.106619.
[5] Das, S. K. (2021). Smart design and its applications: Challenges and techniques. *Nature-Inspired Computing for Smart Application Design*, 1.
[6] Dey, N., Hassanien, A. E., Bhatt, C., Ashour, A., & Satapathy, S. C. (Eds.). (2018). *Internet of Things and Big Data Analytics toward Next-generation Intelligence* (Vol. 35). Springer, Berlin.
[7] Keerthika, M., & Shanmugapriya, D. (2021). Wireless sensor networks: Active and passive attacks-vulnerabilities and countermeasures. *Global Transitions Proceedings*, 2(2), 362–367.
[8] Wan, J., & Chen, J. (2022). AHP based relay selection strategy for energy harvesting wireless sensor networks. *Future Generation Computer Systems*, 128, 36–44.

[9] Dey, N., Samanta, S., Chakraborty, S., Das, A., Chaudhuri, S. S., & Suri, J. S. (2014). Firefly algorithm for optimization of scaling factors during embedding of manifold medical information: An application in ophthalmology imaging. *Journal of Medical Imaging and Health Informatics*, 4(3), 384–394.

[10] Das, S. K., Dao, T. P., & Perumal, T. (Eds.). (2021). *Nature-Inspired Computing for Smart Application Design*. Springer Nature, Singapore.

[11] Misra, Y., Krishnaveni, K., & Rajasekaran, A. S. (2022). Implementation of NLOS based FPGA for distance estimation of elderly using indoor wireless sensor networks. *Materials Today: Proceedings*, https://doi.org/10.1016/j.matpr.2022.01.087.

[12] Wang, F., & Hu, H. (2021). Coverage hole detection method of wireless sensor network based on clustering algorithm. *Measurement*, 179, 109449, https://doi.org/10.1016/j.measurement.2021.109449.

[13] De, D., Mukherjee, A., Das, S. K., & Dey, N. (2020). Wireless sensor network: Applications, challenges, and algorithms. In *Nature Inspired Computing for Wireless Sensor Networks* (pp. 1–18). Springer, Singapore.

[14] Temene, N., Sergiou, C., Georgiou, C., & Vassiliou, V. (2022). A survey on mobility in wireless sensor networks. *Ad Hoc Networks*, 125, 102726, https://doi.org/10.1016/j.adhoc.2021.102726.

[15] Yousefpoor, M. S., Yousefpoor, E., Barati, H., Barati, A., Movaghar, A., & Hosseinzadeh, M. (2021). Secure data aggregation methods and countermeasures against various attacks in wireless sensor networks: A comprehensive review. *Journal of Network and Computer Applications*, 103118, https://doi.org/10.1016/j.jnca.2021.103118.

[16] Das, S. K., Samanta, S., Dey, N., Patel, B. S., & Hassanien, A. E. (Eds.). (2021). *Architectural Wireless Networks Solutions and Security Issues*. Springer, Singapore.

[17] Zhang, J., & Mao, H. (2021). Multi-factor identity authentication protocol and indoor physical exercise identity recognition in wireless sensor network. *Environmental Technology & Innovation*, 101671, https://doi.org/10.1016/j.eti.2021.101671.

[18] Huanan, Z., Suping, X., & Jiannan, W. (2021). Security and application of wireless sensor network. *Procedia Computer Science*, 183, 486–492.

[19] De, D., Mukherjee, A., Das, S. K., & Dey, N. (Eds.). (2020). *Nature Inspired Computing for Wireless Sensor Networks*. Springer, Singapore.

[20] Anand, M., Balaji, N., Bharathiraja, N., & Antonidoss, A. (2021). A controlled framework for reliable multicast routing protocol in mobile ad hoc network. *Materials Today: Proceedings*, https://doi.org/10.1016/j.matpr.2020.10.902.

[21] Tahir, A., Shah, N., Abid, S. A., Khan, W. Z., Bashir, A. K., & Zikria, Y. B. (2021). A three-dimensional clustered peer-to-peer overlay protocol for mobile ad hoc networks. *Computers & Electrical Engineering*, 94, 107364, https://doi.org/10.1016/j.compeleceng.2021.107364.

[22] Das, S. K., Samanta, S., Dey, N., & Kumar, R. (Eds.). (2020). *Design Frameworks for Wireless Networks*. Springer, Singapore.

[23] Sharma, A. S., & Kim, D. S. (2021). Energy efficient multipath ant colony based routing algorithm for mobile ad hoc networks. *Ad Hoc Networks*, 113, 102396, https://doi.org/10.1016/j.adhoc.2020.102396.

[24] Lee, S., Youn, J., & Jung, B. C. (2020). A cooperative phase-steering technique in spectrum sharing-based military mobile ad hoc networks. *ICT Express*, 6(2), 83–86.

[25] Das, S. K., Maheswari, V., & Sharma, A. (2021). Wireless networks: Applications, challenges, and security issues. In *Architectural Wireless Networks Solutions and Security Issues* (pp. 1–10). Springer, Singapore.

[26] Ahmed, H. I., Nasr, A. A., Abdel-Mageid, S. M., & Aslan, H. K. (2021). DADEM: Distributed attack detection model based on big data analytics for the enhancement of the security of internet of things (IoT). *International Journal of Ambient Computing and Intelligence (IJACI)*, 12(1), 114–139.

[27] Sholla, S., Mir, R. N., & Chishti, M. A. (2021). A fuzzy logic-based method for incorporating ethics in the internet of things. *International Journal of Ambient Computing and Intelligence (IJACI)*, 12(3), 98–122.

[28] Balusa, B. C., & Gorai, A. K. (2021). Development of fuzzy pattern recognition model for underground metal mining method selection. *International Journal of Ambient Computing and Intelligence (IJACI)*, 12(4), 64–78.

[29] Singh, A. K., Pamula, R., & Srivastava, G. (2022). An adaptive energy aware DTN-based communication layer for cyber-physical systems. *Sustainable Computing: Informatics and Systems*, 100657, https://doi.org/10.1016/j.suscom.2022.100657.

[30] Singh, A. K., & Pamula, R. (2021). An efficient and intelligent routing strategy for vehicular delay tolerant networks. *Wireless Networks*, 27(1), 383–400.

9 Machine Learning-Based Data Analysis for Managing Challenges of COVID-19
A Survey

Santosh Kumar Das and Joydev Ghosh

CONTENTS

9.1 Introduction 155
9.2 Literature Review 156
9.3 Analyzing Principle of Machine Learning for COVID-19 167
9.4 Conclusions 170
9.5 References 170

9.1 INTRODUCTION

In the last few years, several issues have been created based on the COVID-19 pandemic. It has spread not only in one particular country but all over the world, based on its different variations. It has several phenomena based on its life cycle. It creates several blockages in real-life applications and also in our daily life. The life cycle of the human has become difficult due to the disease, affecting every sector of life, including education, business, services, different organizations, employment, etc. Directly or indirectly, it affects all sectors of our lives. Identification and diagnosis have become difficult due to its variations and several mutations over time. There are also several symptoms available for detection of the disease, such as abdominal pain, chills with shivering, convulsions or seizures, dehydration, issues in breathing, chest pain, general weakness, several types of headaches, irritability, loss of appetite, mental confusion, muscle aches, persistent vomiting, stiff neck, sweating, unusual skin rash, etc. In this article, a survey report is prepared for COVID-19 issues and diagnosis based on the machine learning technique [1–2]. The machine learning technique is a subset of artificial intelligence which is used for analyzing and predicting different types of data available in a website or any repository. It may also be available in physical locations, like in the hospital or any other diagnostic center.

DOI: 10.1201/b23138-12

Several illustrations are maintained based on machine learning techniques based on several applications which are maintained in different survey reports [3–4]. Moreover, several works are proposed based on machine learning techniques for industrial applications, wireless networks, wireless sensor network, and smart application design [5–6].

The remaining part of this article is divided as thus: Section 2 describes a detailed analysis of literature review on COVID-19. Section 3 discusses analysis on principles of machine used in COVID-19. Section 4 is used for the conclusion.

9.2 LITERATURE REVIEW

In the last few years, several research based on machine learning for data analysis and prediction of COVID-19 have been made. In this section, some of the articles are illustrated in a chronological order. Pahar et al. [7] proposed a method for cough analysis and its different types of classification based on COVID-19. The work is based on machine learning algorithm and its different inherent elements. The work also used some of smartphone systems for resolving the issues. The disease is classified into two types; first is normal cough, and second is forced cough. It is based on some dataset used for COVID-19 patients. This article also illustrated that manpower or workload is to be optimized because day by day in that situation, the number of patients increases simultaneously. There are several techniques used in machine learning, such as multilayer perceptron, support vector machine, convolution neural network, k-nearest neighbor algorithm. Finally, in this work, two types of cough are analyzed; one is based on positive case, and the second is on normal case. Kyriazos et al. [8] designed a method for COVID-19 quarantine system based on machine learning. It is based on several well-being score systems. The work is based on a machine learning system for modeling different datasets based on quarantine information available in websites. It is based on Diener's subjective well-being. It is based on factor analysis and management, where data is classify based on information, and it treated 25% as information. Finally, it helps produce several information based on the fusion of machine learning and exploratory graph analysis system. Some of the issues are also based on several wireless network architecture connectivity based on communication systems mentioned in the book [9]. Gulati et al. [10] illustrated a method based on tweet analysis of different information on COVID-19 and its different pandemic variations. The method is based on several types of analyses and information that deal with machine learning techniques. The work is based on comparative analysis, which is done on a sentiment analysis technique which is based on machine learning. The dataset used in this model first is analyzed based on a traditional method, known as lexicon-based approach, then it os extended with help to sentiment analysis. De Fátima Cobre [11] proposed a method based on severity analysis of COVID-19. In this article, several types of predictions and diagnoses are done based on several information. The work is based on biochemical test analysis on machine learning algorithm. It helps analyze different information of the stated issue based on machine learning. The work is a fusion of several types of algorithms, such as artificial neural network, k-nearest neighbor algorithm, and decision tree, to improve the results. Kassania et al. [12] designed a method based on COVID-19 which is part of automatic detection of COVID disease. Two basic components used in this article are CT image and X-ray, and based on these two medical information, disease is detected. A machine learning tool is used to detect the information. Several symptoms used in this model include fever, sore throat, and cough. There are several machine learning algorithms used in this article, such as ResNet, DenseNet, MobileNet, NASNet, etc., for predicting the disease efficiently and effectively. Optimization techniques are used in COVID-19 issue management to help model several wireless network optimization and nature inspired optimization techniques. It helps model several applications and formulations for problem solving [13–14]. Alballa and Al-Turaiki [15] designed a method of severity risk analysis based on different predictions of severity along with mortality and the diagnosis of different diseases based on COVID-19. It is based on machine learning techniques for the analysis of different parameters. The article is based on several types of data on laboratory and clinical information. All information is based on analysis and prediction on different features of prognostic and different types of implementation. Tamal et al. [16] proposed a method of rapid early system based on the diagnosis of different types of X-ray. This X-ray is based on chest analysis and prediction based on different radiomic systems. The work is based on a framework system on different integration of systems. The work is totally based on COVID-19 disease based on different dependent datasets. The images of this analysis are based on pneumonia and normal lung. Z-scoring and bagging model tree are used for this analysis and management. Shahid et al. [17] proposed a method based on virus detection system for COVID-19. It is based on different types of spreading system and analysis based on some factors and information. The work is a fusion of three types of analysis, including medical assistance, virus detection, and spread prevention system. A machine learning tool is used as a prediction and analysis system of different types of modeling. Several types of illness are analyzed based on disease, especially for age sixty or greater. Diagnosis is based on high demand and analysis system based on different types of screening and tracking system. Ebinger et al. [18] designed a method based on algorithm analysis and prediction for COVID-19 patients and its management. The prediction is observed during the time of hospitalization. Different types of variations are measure based on factors and systems of patients' length of stay in the hospital. It is also called LOS based on electronic length analysis of different factors of information. There are three models designed for the purpose of analysis, namely hospital information based on days 1, 2, and 3. The model is analyzed based on some factors of information and its management. Sometimes, some of the issues are modeled based on wireless sensor network application management on different issues and challenges [19]. Shaban et al. [20] designed a patient analysis system used for inference system and design. It is based on several information that help model several neural network systems. The work is based on fuzzy inference modeling system that helps model some issues of COVID-19. The issue is based on deep learning and fuzzy system modeling. It helps model several network information based on strategy management. It helps validate several cross analysis and validation that helps in accuracy modeling. It helps model several detection systems for coronavirus analysis to help in modeling based on a prevention system. Lam et al. [21] designed a model for precision analysis of a model based on medicine analysis and its prescribing system. The model is based on COVID-19 and the different factors of the disease. The information is also collected from pharmacotherapy management based on different factors. In this method, gradient boost tree is used as machine learning model based on several factors and parameters. The algorithm is based on parameters of gradient boost tree based on two basic analyses. First is an indicator based on patients that are in need of and survive with oxygen. Second is based on patients that have survived based on treatment. Weng et al. [22] proposed a method for volatility forecasting system based on online learning system and model. It is based on different factors and information based on a regularization system. The work is based on the COVID-19 pandemic and its different variations models. The method is based on a genetic algorithm on several learning methods. It helps analyze updated learning ability system based on some factors and information. Guzmán-Torres et al. [23] designed a method of analysis of death risk for COVID-19. It is based on COVID-19 patient ratio, which increases rapidly based on several factors. It helps analyze several SARS-CoV-2 conditions based on different parameters and its estimation. All cases are discussed in the model based on the Mexico system. The work is based on different types of data on factors of data science and its analysis. It helps in managing several transmission models for a dynamic system. Some of the factors that influence disease progression include chronic diseases, age, eating habit, and different contacts with infected persons. So in this model, different health conditions are analyzed based on the dynamic parameters of COVID-19. Yao et al. [24] designed a method based on tweet analysis and management of machine learning method for COVID-19. The main purpose of this article os to analyze different information based on sentiment analysis to compare systems of tweet information. Different influence information are managed based on public health and economic analysis for different investigations. The work is based on tweet clustering and management for different months for positive analysis. It also illustrates different correlated information based on different cases and its hospitalization system. Huang et al. [25] proposed a data-driven method for COVID-19 detection. The work is done in Pakistan at Lahore. The method is based on machine learning technique based on different data-driven systems. The work is based on weakness analysis on different variations. The weakness is analyzed based on current test methodology. In this method, priority ranking and logistic regressions are used for analysis purposes and modeling. Local data is analyzed based on concepts of the data-driven method for optimal solution. The main purpose of this solution is to make the system smart based on a design that consists of less issues in terms of challenges and diseases [26]. Tiwari et al. [27] designed a method for vulnerability index based on COVID-19. The method is based on a machine learning technique. The vulnerability analysis is based on death rate of the US based on several types of analysis. It is based on economic, health, and social systems. The community is analyzed based on heterogeneity system of information. In this article, vulnerability is divided into some categories, including low and very low, for analysis purposes. Finally, it helps in public health and different types of disaster management based on requirement analysis. Wu et al. [28] proposed a method for different types of diagnosis of COVID-19 based on machine learning techniques. Machine learning techniques are based on the fusion of texture feature and different types of classifier. In this article, the authors illustrate how COVID-19 differs from pneumonia based on several parameters. The method uses random forest method for analysis of disease based on some factors. Data analysis is based on the correlation system on different images to identify the actual cause of the disease. Pia and Lima [29] designed a method for COVID-19 that helps analyze different suspected patients for COVID based on traditional diagnostic method with the epidemic. In this diagnosis, CT scan is done for chest based on artificial intelligence. There are several machine learning methods used in this article, such as extreme learning with co-occurrence matrix based on gray level. The method is hybrid, based on gray level co-occurrence for image analysis of different classification. Pourhomayoun and Shakibi [30] proposed a method of analysis and predicting mortality based on COVID-19. It is based on several machine learning techniques that help analyze different medical systems. In this article, several types of images are analyzed based on efficient decision-making systems. The proposed method is the fusion of several techniques of machine learning, such as artificial neural network, decision tree, support vector, random forest, logistic regression, and k-nearest neighbor methods. Finally, it helps in depth analysis of the confusion matrix based on different classifiers. Lip et al. [31] designed a method based on risk analysis and prediction of COVID-19 based on patient development. It is based on a machine learning method for the purpose of different algorithms. It is based on multimorbid patient analysis. In this article, several diverse conditions are discussed based on a multimorbid system by changing the methodology of multimorbidity. Finally, it outperforms the results based on parameters and efficiency of clinical testing and validates it. Wang et al. [32] proposed a method based on statistical analysis for the purpose of COVID-19. It is based on acquiring information for outbreak analysis of global pandemic. The model is based on statistical analysis using machine learning. The method uses K means clustering based on a data-driven system for analysis. It is based on model analysis for different patterns. The method is based on the distribution system of vaccines, and it also helps produce optimal results. A. Saygılı [33] designed a method based on machine learning. The purpose of this method is to identify coronavirus based on different image processing systems. It is based on two types of images, namely, X-ray and CT scan. It this method, the authors also illustrate different methods for reducing the spread of the virus based on some parameters, such as hygiene, diet, mask wearing, and maintaining social distance. It this article, the polymerase chain reaction system is used based on image processing to reduce virus spreading. Each information is processed and maintained based on a computerized system. Finally, it helps in dimension reduction, acquisition, feature extraction, and preprocessing system for modeling different information. Arvind et al. [34] designed a machine learning system based on predicting intubation system on COVID-19. It analyzed several information based on several factors and information, such as demographic data, vital information management, and laboratory information system. It is based on different information and management of different systems. The method helps predict several factors and information that help in managing several parameters. It helps in managing and reducing the spread of COVID-19. Choudrie et al. [35] proposed a method based on online information system of misinformation based on different factors. The information is based on COVID-19 analysis. It is used to design several information based on some of information from different systems. It is based on different machine learning systems that help different methods by collecting information. Khan et al. [36] proposed a method for projecting several information based on a critical analysis of COVID-19. It is based on the fusion of machine learning and GIS system based on different transmission of the model. It is based on some processes of different operations, such as provision of lockdown system, thermal screening system, sanitizer use, and social distance maintenance, and quarantines. It is based on different factors and information. Several methods of machine learning are used in this system based on the fusion of decision tree, Gaussian method for process regression, and support vector machine. Finally, it outperforms the result based on different metrics. Alves et al. [37] designed a method based on different diagnostic systems based on machine learning method. It is used for COVID-19 analysis based on blood testing. This testing is based on several examinations for the identification of COVID-19. Several machine learning tests are used for different purposes based on the fusion of decision tree based on local graph analysis and different classifier systems. It is based on interconnection system for factor analysis. It is based on electronic health records for analysis. Magazzino et al. [38] designed a nexus-based analysis for COVID-19 for death analysis. The analysis is done based on factors of machine learning used by deep learning. The research is based on the state of New York. This study is based on several factors, such as economic growth, fine particular, and nitrogen dioxide system. Finally, it helps reduce several types of pollution based on the reduction of COVID-19 analysis and management. Numerical analysis is based on a machine learning system based on different factors. Dairi et al. [39] proposed a method for comparative analysis of COVID-19 based on different forecasting methods. It is based on different types of transmission for different systems of forecasting. It is based on different system models and innovations based on different factors. The work is based on the convolution neural network with the help of restricted Boltzmann machine system for the support vector machine. The analysis is based on different system modeling based on the countries of Mexico, Brazil, France, Russia, Saudi Arabia, and also the US. Kabir and Madria [40] designed an emotion system for visualization of data and information on COVID-19. It used emotion detection for tweet analysis and preservation. All information is done by the model based on factors and information available in the tweet. It is based on emotion analysis for resenting positive emotion, negative emotion, and neutral emotion used by people. The main tool used in this method is machine learning based on several factors of classification analysis. Saha et al. [41] designed a method for COVID-19 for different types of diagnostic purposes for the doctor and the patient. It is based on convolution neural network, which is part of the machine learning system. It is based on the convolution of neural network based on different factors. It is based on several methods based on several factors and information. It helps in managing several operations based on X-ray images with the help of several factors, such as random forest analysis, decision tree, machine classifier, and AdaBoost system. Finally, it helps manage several methods to outperform the result of the model for efficiency. Quintero et al. [42] proposed a method for prediction of variable analysis of different models. It is based on a deep learning model for handling different purposes of COVID-19. It is based on different recovering systems and models for analysis of variable. It is based on intradependence and contextual analysis. The method is based on persons whose age is greater than 65. Based on several prediction and analysis, it helps guide the model efficiently and effectively based on different variations. Satu et al. [43] designed a method of machine learning for COVID-19 tweet analysis. It is based on investigation of classification system. It helps in analysis based on machine learning method for the analysis of different types of data and information. The method is based on several types of data analysis for the purpose of symptomatic analysis system. It is based on several information on different influence and index mechanisms. It helps enhance several information based on dataset mechanism. Chakraborty and Mali [44] designed a framework of COVID-19 analysis and management based on different types of image processing and information. It is based on data interpretation on different factors and information. The technique is based on image processing for the purpose of learning. The method is robust and uses radiological analysis to help the system. The X-ray of this method is based on a heuristic method that uses two heuristic methods using fuzzy logic and cuckoo search analysis. Praveen et al. [45] proposed a method of tweet analysis and management. It is based on the analysis of 840,000 tweet data on COVID-19. It is a fusion of several information or parameters, such as trauma, anxiety, and stress for handling different information. Data is based on social media information on Indians. It also includes several reports based on actual death rate on the system. Pinheiro et al. [46] designed a method based on network analysis and management to identify different information. This information is based on virus analysis and management to identify different trends that are spreading. It is based on COVID-19 information that helps in managing several information for detection. Machine learning techniques help in managing different information using location analysis of network based on different topologies. Gothai et al. [47] designed a method based on COVID-19 analysis for machine learning. It is based on machine learning for different trend analyses. It helps in X-ray analysis of different systems. It is based on model analysis for the support vector analysis by using regression. It is based on different machine learning methods to analyze system and design. It uses different types of information for the classification of support vector machine. It is based on different types of system analyses and design for COVID-19 analysis. Tianqing et al. [48] proposed a method for real-time analysis on COVID-19. This is based on image processing for X-ray analysis. It is based on extreme analysis for optimization. It helps analyze data based on convolution neural network analysis. It is based on whale optimization system for analyzing the method. It is based on comparative analysis based on benchmark analysis. There are several heuristic algorithms used in this system, such as cuckoo search, whale optimization, genetic algorithm, etc. It helps analyze the results. P. Sv et al. [49] designed a method for detection of COVID-19 side effect for diagnosis based on different information. It is based on Indian citizen analysis on different perspective systems. The work is based on a machine learning system that uses several vaccines based on two points of views; first is neutral, and second is positive. Each data or dataset is based on social media information for patient analysis. It helps several Indian persons' attitudes. Sentiment analysis is used based on different positive and negative aspects of information. S. Ballı [50] proposed a method of data analysis for handling the COVID-19 pandemic. It is based on short-term analysis and prediction system for handling different forecasting methods. The method is based on time series analysis and management. It is based on a cumulative method for handling different information. Several coronavirus cases are analyzed based on a prediction system in the country of the USA as well as Germany for a more global point of view. The method uses support vector machine for performance analysis based on a machine learning system. It helps deal with different metrics based on statistical analysis and predictions. M. Turkoglu [51] designed a COVID-19 analysis of different CT scan images of the chest for multiple kernel system. It is based on extreme learning method that uses the fusion of deep learning and machine learning. CT scan is based on different phenomena, such as normal as well as pneumonia. This is based on data analysis and prediction for classification analysis and accuracy management based on more than one kernel system. Finally, it helps analyze several predictions based on the identification of different diseases. Chakraborti et al. [52] explored several information on determinant analysis based on different scenarios of the pandemic. It is based on plausible application on information of COVID-19 analysis. The model is based on epidemiological analysis that uses different methods of machine learning, like gradient-boosted algorithm and random forest. It deals with different air pollution sources based on the relative influence system. There are several variables used in this model, such as economy, air pollution, demographic, migration, etc., based on several controlling system. It helps design several superiority model analysis and predictions. S. Cerna et al. [53] proposed a firemen ambulance system for turnaround analysis. It is based on time prediction for handling different hospitals. It is based on COVID-19 for different emergency medical services and predictions. It helps analyze different types of services and several breakdown of services. The machine learning method helps propose and analyze several weather and statistical data analysis and prediction system. Several fusions of machine learning techniques are used as multilayer perceptron, prophet, light gradient-boosted machine, and long short-term memory analysis system. The work is based on a multivariate model of analysis for handling different decision-making systems. Tarik et al. [54] proposed a method of prediction and analysis of performance of students. This prediction is based on the COVID-19 pandemic, using the fusion of two methods, namely, machine learning and artificial intelligence. It helps map several enabling machines for system and information. It helps design an efficient decision-making system for an intelligent design. The performance is analyzed in students of Moroccan descent in Guelmim Oued Noun. It helps improve several information based on different experience analyses. Zhou et al. [55] designed a method of routine management of clinical feature analysis and prediction. It is based on uncovered severity analysis on COVID-19. The technique is based on a machine learning system for prediction. The technique of machine learning is used for longitudinal measurement system. It is based on a data matrix that consists of several matrix elements, such as different types of measurement, reading information based on disease progression. It is based on classifier analysis for different types of values, such as predictive value, sensitivity value, specificity value, based on dynamic analysis. This is based on normal range analysis and prediction system. D. Yacchirema and A. Chura [56] proposed a method of safe mobility analysis and management. The work is based on the fusion of internet of things and machine learning system for handling different types of predictions. All the information is based on the format of age of COVID-19 analysis. There are several options mentioned in this model for reducing different issues of the pandemic. Some of the factors used in this model are multilayer perceptron, dataset analysis, machine learning, fog node analysis, and different real-life analyses. Zivkovic et al. [57] designed a purpose of modeling for hybrid machine learning techniques. It is based on the beetle antennae search approach system. The method is based on forecasting approach for identifying several cases of the system. The method uses the fusion of neural network and fuzzy logic. It is based on an enhanced algorithm for different substantial systems. The method is used to compare several models for managing several points and information systems. The method is based on influenza analysis to confirm different cases in the USA and China for simulation analysis. Mojjada et al. [58] designed a model for future analysis-based forecasting system. It is based on a machine learning technique for analysis of disease. There are several operators used in this model, such as linear regression, selection operator, and lower absolute redactor. The method used in this article is analyzed based on several parameters on exponential analysis. It also uses support vector analysis for lower absolute analysis. In this article, different death rates are analyzed based on recovery also. Burdick et al. [59] proposed a method based on respiratory prediction for different decompensation systems. The method is based on machine learning and prediction system for the analysis of different patients. The work is based on ready trial basis for handling different issues. It helps in managing several types of prognosis systems based on accurate methods of alternative strategy management. It handles several types of risk based on different factors and information management. Malki et al. [60] designed a method for different association of modeling for weather forecasting and analysis. It is based on pandemic analysis for different mortality systems. It helps design several mortality rates of the system. The method is based on machine learning system that uses different seasonal pattern analysis and management. The pattern of this method helps design based on several modeling and analysis on different distribution systems. For handling this system, several information, like humidity, temperature, and different transmission, are used to manage for different variable analysis and management. Dandekar et al. [61] proposed a machine learning method of global diagnosis and analysis based on comparative tools and management. It is based on quarantine information managing and controlling. Each of the information is based on the spread of COVID-19. It helps in managing different assessments of information and its effect on the pandemic. It uses several intersection methods for machine learning. It helps in managing several growths of the system and varies based on the measurement of quarantine system and management. The controlling method is based on data of different countries, such as those in Asia, Europe, South America, and North America, based on different metrics and economic management. Doanvo et al. [62] designed a method for map analysis and management. It is based on several research based on requirement system. The work is based on machine learning system and management. It also used artificial intelligence technique for the main goal, which is prioritization system. It helps in managing several pandemic information to measure different resource allocation systems and management. Shaban et al. [63] designed a patient analysis system that uses inference system and design. It is based on several information that help model several neural network systems and modeling. The work is based on fuzzy inference modeling that helps model some issues of COVID-19. The issue is based on deep learning and fuzzy system modeling. It helps model several network information based on strategy management. It helps validate several cross analysis and validation that help in accuracy. It helps model several detections for coronavirus to help make a model based on a prevention system. Zahra et al. [64] designed a method for data-driven system that is based on intelligence system maintenance and information system. The work is based on URL system of information and modeling that helps analyze and design several malicious and phishing information system. The work is completely based on fuzzy logic modeling. It helps deal with several uncertainty management for handling pandemic analysis. It helps model several control and unprecedented information systems. The work helps model cybercriminal information analyses to deal with several ransomware information systems for impact analysis. It uses an efficient decision-making system to mitigate different community information based on different discrepancies [65–67]. In the aforementioned literature, several information is discussed in the form of new innovation and research for handling critical analysis of the disease based on different variations. Its summarized information is shown in Table 9.1, and its related abbreviations are shown in Table 9.2.

TABLE 9.1

Summarized Information of Literature

Ref.	Year	Purpose	Used Technique
[7]	2021	Cough Classification	Logistic Regression, KNN, SVM, MLP, CNN, LSTM
[8]	2021	Quarantine	EGA
[10]	2021	Tweets Analysis for COVID-19	Sentiment Analysis, Classification
[11]	2021	Diagnosis and Prediction Severity	Decision Tree, ANN, PLS-DA, KNN
[12]	2021	Automatic Detection of Virus	Classification, CNN
[15]	2021	Diagnosis, Mortality, and Severity Risk Prediction	Machine Learning, Training Dataset
[16]	2021	Rapid Early Diagnosis by Chest X-Ray	ANOVA Test, SVM, KNN, EBM
[17]	2021	Virus Detection and Spread Prevention	Machine Learning
[18]	2021	Predicts Duration of Hospitalization	Machine Learning, EHR
[20]	2021	COVID Patient Detection	FL and ANN
[21]	2021	Prescribing COVID-19 Pharmacotherapy	GBDT
[22]	2021	Volatility Forecasting of Crude Oil Futures	Genetic Algorithm
[23]	2021	Estimation of Conditions for Patients Due to Cause of Death	Machine Learning
[24]	2021	Comparing Tweet Sentiments	Machine Learning, Sentiment Analysis
[25]	2021	Socioeconomic Planning	Machine Learning, Data-Driven Test
[27]	2021	Vulnerability Index Measurement	Machine Learning
[28]	2021	Texture Feature-Based Diagnosis	RFC
[29]	2021	COVID Diagnosis by Chest CT	Co-occurrence Matrix, ELM
[30]	2021	Predicting Risk of Patients	Logistic Regression, Random Forest, Decision Tree, ANN, KNN
[31]	2021	Risk Prediction in Patients	Machine Learning
[32]	2021	Global Transmission	Machine Learning, Classification, Statistical Analysis
[33]	2021	Coronavirus Detection by CT and X-Ray Images	Machine Learning, Image Processing, Classification
[34]	2021	Predict Intubation in Hospitalized Patients	Machine Learning
[35]	2021	Online Information Processing	Artificial Intelligence, Machine Learning, CNN
[36]	2021	Projecting Criticality Situation of COVID	Machine Learning, SVM, GPR
[37]	2021	Blood Test–Based COVID Diagnosis	MLC, DTX

(Continued)

TABLE 9.1   (Continued)

Ref.	Year	Purpose	Used Technique
[38]	2021	Death Analysis in New York by COVID	DML
[39]	2021	Transmission Forecasting of COVID-19	Logistic Regression, LSTM-CNN, RBM, SVR
[40]	2021	Emotion Visualization, Analysis, and Detection	Classification, ANN
[41]	2021	X-Ray Image for COVID-19 Analyzing	CNN and MLC
[42]	2021	SEIRD Analysis	Machine Learning, Predicting Model
[43]	2021	Tweets Analysis	Clustering, MLC
[44]	2021	Image Segmentation for Radiological Image	FL, Cuckoo Search, Image Processing
[45]	2021	Analyzing Stress, Trauma, and Anxiety	Machine Learning
[46]	2021	Checking Spread of Virus	Machine Learning
[47]	2021	Growth and Trend Analysis of COVID	SML
[48]	2021	COVID Analysis Using X-Ray	CNN, ELM, COA,
[49]	2021	Side Effect Analysis for COVID-19 Vaccine	Machine Learning, Sentimental Analysis
[50]	2021	Data Analysis for COVID-19	Machine Learning, Statistical Distributions, SVM
[51]	2021	COVID Analysis Using Chest CT Images	ELM, DNN
[52]	2021	Determinant Factor Analysis of COVID	Machine Learning, Random Forest, GBM
[53]	2021	Ambulance Turnaround Time Analysis in Emergency Situation	Multilayer Perceptron, LGBM, LSTM
[54]	2021	Predicting Performance of Student	Artificial Intelligence, Machine Learning
[55]	2021	Predicting Features of Clinical Data	Machine Learning, PPV, NPV
[56]	2021	Mobility Based on Safety	Machine Learning, IoT
[57]	2021	Case Analysis of COVID-19	Machine Learning, Swarm Intelligence, ANFIS
[58]	2020	Future Analysis of COVID-19	Machine Learning, Logistic Regression, SVM, LASSo
[59]	2020	Decompensation of Respiratory Prediction	Machine Learning, MEWS
[60]	2020	Weather Data Analysis and Prediction	Machine Learning
[61]	2020	Global Diagnostic for Quarantine	Machine Learning
[62]	2020	Literature Analysis	Artificial Intelligence, Machine Learning
[63]	2021	COVID-19 Detection	FL
[64]	2021	Detecting COVID-19 along with Attack	FL

TABLE 9.2
List of Abbreviation with Description

Abbreviation	Description
ANFIS	Adaptive Neuro-Fuzzy Inference System
ANN	Artificial Neural Networks
CNN	Convolutional Neural Network
COA	Chimp Optimization Algorithm
DML	Deep Machine Learning
DNN	Deep Neural Network
DTX	Decision Tree Explainer
EBM	Ensemble Bagged Model
EGA	Exploratory Graph Analysis
EHR	Electronic Health Record
ELM	Extreme Learning Machine
FL	Fuzzy Logic
GBDT	Gradient-Boosted Decision Tree
GBM	Gradient-Boosted Machine
GPR	Gaussian Process Regression
IoT	Internet of Things
KNN	k-Nearest Neighbor
LASSo	Lower Absolute Reductor and Selection Operator
LGBM	Light Gradient-Boosted Machine
LNA	Location Network Analysis
LSTM-CNN	Long Short-Term Memory-Convolutional Neural Networks
MEWS	Modified Early Warning Score
MLC	Machine Learning Classifier
MLP	Multilayer Perceptron
NPV	Negative Predictive Value
PLS-DA	Partial Least Squares Discriminant Analysis
PPV	Positive Predictive Value
RBM	Restricted Boltzmann Machine
RFC	Random Forest Classifier
SML	Supervised Machine Learning
SVM	Support Vector Machine
SVR	Support Vector Regression

9.3 ANALYZING PRINCIPLE OF MACHINE LEARNING FOR COVID-19

The paper is based on several issues and their analyses based on machine learning technique. All the stated issues are based on a new disease, i.e., COVID-19, and its different variations. It helps give new ideas in the field of research and innovation. Machine learning is a part of soft computing as well as artificial intelligence for calculation and analysis of data and information. It is the working principle shown in Figure 9.1. It helps in managing several information based on analysis of data with information. Because there are several types of data available based on nature, such as high-dimensional data, machine data, operational data, spatio-temporal data, time-stamped data, unverified outdated data, etc., in some cases, data is categorized based on real-time data, such as dark data and open data. Day by day, the size of data increases rapidly, so the concept of big data rises in the area of research. It works with cloud computing and IoT, or internet of things. Each data is based on two basic phases, namely, training, then testing. The working principle of machine learning is based on, first, the need to collect data and then analyze it based on several variations. Then check it for the purpose of suitability and make it efficiently. If it's suitable, then good; otherwise, it's repeated

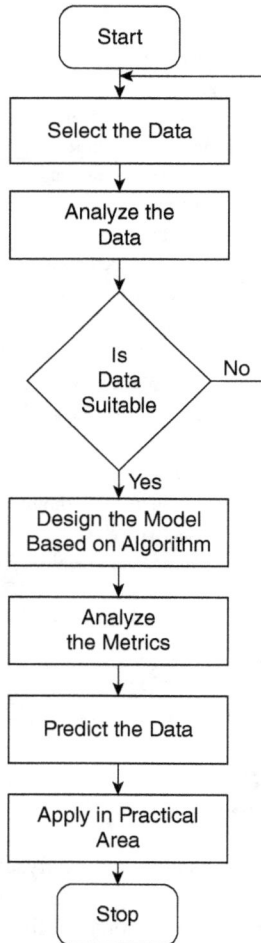

FIGURE 9.1 Flowchart of the working principle of machine learning.

from the phase of data selection. After checking data suitability, then model for analysis based on the proposed algorithm and method. For the purpose of testing and validating, analyze different metrics for better outcomes. Finally, predict the data and apply it in real-life area.

Figure 9.2 indicates a block diagram of the machine learning process for data analysis and design system. It consists of several phases and conditions for handling several types of data that are present in real-life areas. So first, select the

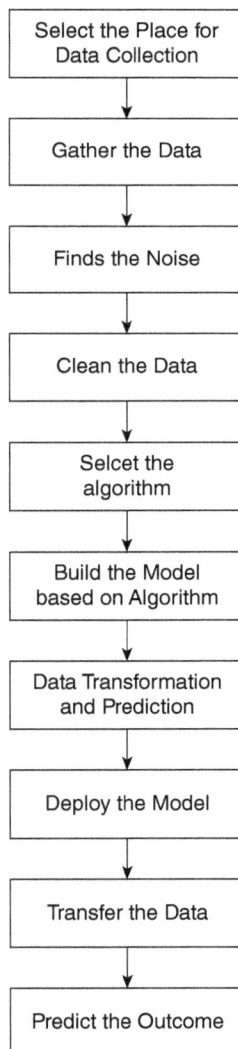

FIGURE 9.2 Block diagram of machine learning analyzing phases.

actual data based on any website or physical place, gather the data for the purpose of finding different types of errors and noises, then clean the data to reduce uncertainty and make precise. For the purpose of modeling, select the algorithm, then build the model of proposed problem-solving. It helps in data transformation and prediction system for handling the deployment of the model. It helps transfer the data based on several types of predictions and producing the actual outcomes.

9.4 CONCLUSIONS

The paper is based on several types of information and its analysis based on COVID-19 and its different variations. It also helps describe several symptoms and analysis of information of new disease to help map several information based on different parameters. It helps in managing several issues based on factors and paradigms of diseases. The overall literature is based on the concept of machine learning and prediction that efficiently helps in the areas of research. This research is based on different algorithms of machine learning. It shows how these algorithms are used to solve issues of any disease based on diagnosis, analysis, prediction, testing, variation, etc. and its management. The parameters used in this system include different types of images, scanning, medicine, quarantine system, maintaining distance, wearing mask, regularly using sanitizer, practicing safety, etc. The information that is available in this paper helps the reader gain a new insight in the area of research and innovation.

9.5 REFERENCES

[1] Garg, A., & Mago, V. (2021). Role of machine learning in medical research: A survey. *Computer Science Review*, 40, 100370, https://doi.org/10.1016/j.cosrev.2021.100370.
[2] Das, S. K., Das, S. P., Dey, N., & Hassanien, A. E. (Eds.). (2021). *Machine Learning Algorithms for Industrial Applications*. Springer, https://doi.org/10.1007/978-3-030-50641-4.
[3] Gambella, C., Ghaddar, B., & Naoum-Sawaya, J. (2020). Optimization problems for machine learning: A survey. *European Journal of Operational Research*, https://doi.org/10.1016/j.ejor.2020.08.045.
[4] Balaji, T. K., Annavarapu, C. S. R., & Bablani, A. (2021). Machine learning algorithms for social media analysis: A survey. *Computer Science Review*, 40, 100395, https://doi.org/10.1016/j.cosrev.2021.100395.
[5] De, D., Mukherjee, A., Das, S. K., & Dey, N. (Eds.). (2020). *Nature Inspired Computing for Wireless Sensor Networks*. Springer, https://doi.org/10.1007/978-981-15-2125-6.
[6] Das, S. K., Dao, T. P., & Perumal, T. (Eds.). (2021). *Nature-Inspired Computing for Smart Application Design*. Springer Nature, https://doi.org/10.1007/978-981-33-6195-9.
[7] Pahar, M., Klopper, M., Warren, R., & Niesler, T. (2021). COVID-19 cough classification using machine learning and global smartphone recordings. *Computers in Biology and Medicine*, 104572, https://doi.org/10.1016/j.compbiomed.2021.104572.
[8] Kyriazos, T., Galanakis, M., Karakasidou, E., & Stalikas, A. (2021). Early COVID-19 quarantine: A machine learning approach to model what differentiated the top 25% well-being scorers. *Personality and Individual Differences*, 181, 110980, https://doi.org/10.1016/j.paid.2021.110980.

[9] Das, S. K., Samanta, S., Dey, N., Patel, B. S., & Hassanien, A. E. (Eds.). (2021). *Architectural Wireless Networks Solutions and Security Issues.* Springer, https://doi.org/10.1007/978-981-16-0386-0.

[10] Gulati, K., Kumar, S. S., Boddu, R. S. K., Sarvakar, K., Sharma, D. K., & Nomani, M. Z. M. (2021). Comparative analysis of machine learning-based classification models using sentiment classification of tweets related to COVID-19 pandemic. *Materials Today: Proceedings*, https://doi.org/10.1016/j.matpr.2021.04.364.

[11] de Fátima Cobre, A., Stremel, D. P., Noleto, G. R., Fachi, M. M., Surek, M., Wiens, A., & Pontarolo, R. (2021). Diagnosis and prediction of COVID-19 severity: Can biochemical tests and machine learning be used as prognostic indicators? *Computers in Biology and Medicine*, 104531.

[12] Kassania, S. H., Kassanib, P. H., Wesolowskic, M. J., Schneidera, K. A., & Detersa, R. (2021). Automatic detection of coronavirus disease (COVID-19) in X-ray and CT images: A machine learning based approach. *Biocybernetics and Biomedical Engineering*, 41(3), 867–879.

[13] Das, S. K., Maheswari, V., & Sharma, A. (2021). Wireless networks: Applications, challenges, and security issues. In *Architectural Wireless Networks Solutions and Security Issues* (pp. 1–10). Springer, Singapore.

[14] Dey, N., Samanta, S., Chakraborty, S., Das, A., Chaudhuri, S. S., & Suri, J. S. (2014). Firefly algorithm for optimization of scaling factors during embedding of manifold medical information: An application in ophthalmology imaging. *Journal of Medical Imaging and Health Informatics*, 4(3), 384–394.

[15] Alballa, N., & Al-Turaiki, I. (2021). Machine learning approaches in COVID-19 diagnosis, mortality, and severity risk prediction: A review. *Informatics in Medicine Unlocked*, 100564, https://doi.org/10.1016/j.imu.2021.100564.

[16] Tamal, M., Alshammari, M., Alabdullah, M., Hourani, R., Alola, H. A., & Hegazi, T. M. (2021). An integrated framework with machine learning and radiomics for accurate and rapid early diagnosis of COVID-19 from chest X-ray. *Expert Systems with Applications*, 180, 115152, https://doi.org/10.1016/j.eswa.2021.115152.

[17] Shahid, O., Nasajpour, M., Pouriyeh, S., Parizi, R. M., Han, M., Valero, M., & Sheng, Q. Z. (2021). Machine learning research towards combating COVID-19: Virus detection, spread prevention, and medical assistance. *Journal of Biomedical Informatics*, 117, 103751, https://doi.org/10.1016/j.jbi.2021.103751.

[18] Ebinger, J., Wells, M., Ouyang, D., Davis, T., Kaufman, N., Cheng, S., & Chugh, S. (2021). A machine learning algorithm predicts duration of hospitalization in COVID-19 patients. *Intelligence-based Medicine*, 100035, https://doi.org/10.1016/j.ibmed.2021.100035.

[19] De, D., Mukherjee, A., Das, S. K., & Dey, N. (2020). Wireless sensor network: Applications, challenges, and algorithms. In *Nature Inspired Computing for Wireless Sensor Networks* (pp. 1–18). Springer, Singapore.

[20] Shaban, W. M., Rabie, A. H., Saleh, A. I., & Abo-Elsoud, M. A. (2021). Detecting COVID-19 patients based on fuzzy inference engine and deep neural network. *Applied Soft Computing*, 99, 106906, https://doi.org/10.1016/j.asoc.2020.106906.

[21] Lam, C., Siefkas, A., Zelin, N. S., Barnes, G., Dellinger, R. P., Vincent, J. L., & Das, R. (2021). Machine learning as a precision-medicine approach to prescribing COVID-19 pharmacotherapy with remdesivir or corticosteroids. *Clinical Therapeutics*, https://doi.org/10.1016/j.clinthera.2021.03.016

[22] Weng, F., Zhang, H., & Yang, C. (2021). Volatility forecasting of crude oil futures based on a genetic algorithm regularization online extreme learning machine with a forgetting factor: The role of news during the COVID-19 pandemic. *Resources Policy*, 73, 102148, https://doi.org/10.1016/j.resourpol.2021.102148.

[23] Guzmán-Torres, J. A., Alonso-Guzmán, E. M., Domínguez-Mota, F. J., & Tinoco-Guerrero, G. (2021). Estimation of the main conditions in (SARS-CoV-2) COVID-19 patients that cause their death using Machine learning, the case of Mexico. *Results in Physics*, 104483, https://doi.org/10.1016/j.rinp.2021.104483.

[24] Yao, Z., Yang, J., Liu, J., Keith, M., & Guan, C. (2021). Comparing tweet sentiments in megacities using machine learning techniques: In the midst of COVID-19. *Cities*, 116, 103273, https://doi.org/10.1016/j.cities.2021.103273.

[25] Huang, C., Wang, M., Rafaqat, W., Shabbir, S., Lian, L., Zhang, J., & Song, W. (2021). Data-driven test strategy for COVID-19 using machine learning: A study in Lahore, Pakistan. *Socio-economic Planning Sciences*, 101091, https://doi.org/10.1016/j.seps.2021.101091.

[26] Das, S. K. (2021). Smart design and its applications: Challenges and techniques. *Nature-Inspired Computing for Smart Application Design*, 1.

[27] Tiwari, A., Dadhania, A. V., Ragunathrao, V. A. B., & Oliveira, E. R. (2021). Using machine learning to develop a novel COVID-19 vulnerability Index (C19VI). *Science of The Total Environment*, 773, 145650, https://doi.org/10.1016/j.scitotenv.2021.145650.

[28] Wu, Z., Li, L., Jin, R., Liang, L., Hu, Z., Tao, L., & Guo, X. (2021). Texture feature-based machine learning classifier could assist in the diagnosis of COVID-19. *European Journal of Radiology*, 137, 109602, https://doi.org/10.1016/j.ejrad.2021.109602.

[29] Pi, P., & Lima, D. (2021). Gray level co-occurrence matrix and extreme learning machine for COVID-19 diagnosis. *International Journal of Cognitive Computing in Engineering*, https://doi.org/10.1016/j.ijcce.2021.05.001

[30] Pourhomayoun, M., & Shakibi, M. (2021). Predicting mortality risk in patients with COVID-19 using machine learning to help medical decision-making. *Smart Health*, 20, 100178, https://doi.org/10.1016/j.smhl.2020.100178.

[31] Lip, G. Y., Genaidy, A., Tran, G., Marroquin, P., & Estes, C. (2021). Incident atrial fibrillation and its risk prediction in patients developing COVID-19: A machine learning based algorithm approach. *European Journal of Internal Medicine*, https://doi.org/10.1016/j.ejim.2021.04.023.

[32] Wang, W. C., Lin, T. Y., Chiu, S. Y. H., Chen, C. N., Sarakarn, P., Ibrahim, M., & Yeh, Y. P. (2021). Classification of community-acquired outbreaks for the global transmission of COVID-19: Machine learning and statistical model analysis. *Journal of the Formosan Medical Association*, https://doi.org/10.1016/j.jfma.2021.05.010.

[33] Saygılı, A. (2021). A new approach for computer-aided detection of coronavirus (COVID-19) from CT and X-ray images using machine learning methods. *Applied Soft Computing*, 105, 107323, https://doi.org/10.1016/j.asoc.2021.107323.

[34] Arvind, V., Kim, J. S., Cho, B. H., Geng, E., & Cho, S. K. (2021). Development of a machine learning algorithm to predict intubation among hospitalized patients with COVID-19. *Journal of Critical Care*, 62, 25–30.

[35] Choudrie, J., Banerjee, S., Kotecha, K., Walambe, R., Karende, H., & Ameta, J. (2021). Machine learning techniques and older adults processing of online information and misinformation: A COVID 19 study. *Computers in Human Behavior*, 119, 106716, https://doi.org/10.1016/j.chb.2021.106716.

[36] Khan, F. M., Kumar, A., Puppala, H., Kumar, G., & Gupta, R. (2021). Projecting the criticality of COVID-19 transmission in India using GIS and machine learning methods. *Journal of Safety Science and Resilience*, https://doi.org/10.1016/j.jnlssr.2021.05.001.

[37] Alves, M. A., Castro, G. Z., Oliveira, B. A. S., Ferreira, L. A., Ramírez, J. A., Silva, R., & Guimarães, F. G. (2021). Explaining machine learning based diagnosis of COVID-19 from routine blood tests with decision trees and criteria graphs. *Computers in Biology and Medicine*, 132, 104335, https://doi.org/10.1016/j.compbiomed.2021.104335.

[38] Magazzino, C., Mele, M., & Sarkodie, S. A. (2021). The nexus between COVID-19 deaths, air pollution and economic growth in New York state: Evidence from deep machine learning. *Journal of Environmental Management*, 286, 112241, https://doi.org/10.1016/j.jenvman.2021.112241.

[39] Dairi, A., Harrou, F., Zeroual, A., Hittawe, M. M., & Sun, Y. (2021). Comparative study of machine learning methods for COVID-19 transmission forecasting. *Journal of Biomedical Informatics*, 118, 103791, https://doi.org/10.1016/j.jbi.2021.103791.

[40] Kabir, M. Y., & Madria, S. (2021). EMOCOV: Machine learning for emotion detection, analysis and visualization using COVID-19 tweets. *Online Social Networks and Media*, 23, 100135, https://doi.org/10.1016/j.osnem.2021.100135.

[41] Saha, P., Sadi, M. S., & Islam, M. M. (2021). EMCNet: Automated COVID-19 diagnosis from X-ray images using convolutional neural network and ensemble of machine learning classifiers. *Informatics in Medicine Unlocked*, 22, 100505, https://doi.org/10.1016/j.imu.2020.100505.

[42] Quintero, Y., Ardila, D., Camargo, E., Rivas, F., & Aguilar, J. (2021). Machine learning models for the prediction of the SEIRD variables for the COVID-19 pandemic based on a deep dependence analysis of variables. *Computers in Biology and Medicine*, 104500, https://doi.org/10.1016/j.compbiomed.2021.104500.

[43] Satu, M. S., Khan, M. I., Mahmud, M., Uddin, S., Summers, M. A., Quinn, J. M., & Moni, M. A. (2021). TClustVID: A novel machine learning classification model to investigate topics and sentiment in COVID-19 tweets. *Knowledge-Based Systems*, 226, 107126, https://doi.org/10.1016/j.knosys.2021.107126.

[44] Chakraborty, S., & Mali, K. (2021). SUFMACS: A machine learning-based robust image segmentation framework for COVID-19 radiological image interpretation. *Expert Systems with Applications*, 178, 115069, https://doi.org/10.1016/j.eswa.2021.115069.

[45] Praveen, S. V., Ittamalla, R., & Deepak, G. (2021). Analyzing Indian general public's perspective on anxiety, stress and trauma during COVID-19-a machine learning study of 840,000 tweets. *Diabetes & Metabolic Syndrome: Clinical Research & Reviews*, 15(3), 667–671.

[46] Pinheiro, C. A. R., Galati, M., Summerville, N., & Lambrecht, M. (2021). Using network analysis and machine learning to identify virus spread trends in COVID-19. *Big Data Research*, 100242, https://doi.org/10.1016/j.bdr.2021.100242.

[47] Gothai, E., Thamilselvan, R., Rajalaxmi, R. R., Sadana, R. M., Ragavi, A., & Sakthivel, R. (2021). Prediction of COVID-19 growth and trend using machine learning approach. *Materials Today: Proceedings*, https://doi.org/10.1016/j.matpr.2021.04.051.

[48] Hu, T., Khishe, M., Mohammadi, M., Parvizi, G. R., Karim, S. H. T., & Rashid, T. A. (2021). Real-time COVID-19 diagnosis from X-Ray images using deep CNN and extreme learning machines stabilized by chimp optimization algorithm. *Biomedical Signal Processing and Control*, 68, 102764, https://doi.org/10.1016/j.bspc.2021.102764.

[49] Sv, P., Tandon, J., & Hinduja, H. (2021). Indian citizen's perspective about side effects of COVID-19 vaccine—A machine learning study. *Diabetes & Metabolic Syndrome: Clinical Research & Reviews*, https://doi.org/10.1016/j.dsx.2021.06.009.

[50] Ballı, S. (2021). Data analysis of COVID-19 pandemic and short-term cumulative case forecasting using machine learning time series methods. *Chaos, Solitons & Fractals*, 142, 110512, https://doi.org/10.1016/j.chaos.2020.110512.

[51] Turkoglu, M. (2021). COVID-19 detection system using chest CT images and multiple kernels-extreme learning machine based on deep neural network. *IRBM*, https://doi.org/10.1016/j.irbm.2021.01.004.

[52] Chakraborti, S., Maiti, A., Pramanik, S., Sannigrahi, S., Pilla, F., Banerjee, A., & Das, D. N. (2021). Evaluating the plausible application of advanced machine learnings in exploring determinant factors of present pandemic: A case for continent specific COVID-19 analysis. *Science of the Total Environment*, 765, 142723, https://doi.org/10.1016/j.scitotenv.2020.142723.

[53] Cerna, S., Arcolezi, H. H., Guyeux, C., Royer-Fey, G., & Chevallier, C. (2021). Machine learning-based forecasting of firemen ambulances' turnaround time in hospitals, considering the COVID-19 impact. *Applied Soft Computing*, 107561, https://doi.org/10.1016/j.asoc.2021.107561.

[54] Tarik, A., Aissa, H., & Yousef, F. (2021). Artificial intelligence and machine learning to predict student performance during the COVID-19. *Procedia Computer Science*, 184, 835–840.

[55] Zhou, K., Sun, Y., Li, L., Zang, Z., Wang, J., Li, J., & Guo, T. (2021). Eleven routine clinical features predict COVID-19 severity uncovered by machine learning of longitudinal measurements. *Computational and Structural Biotechnology Journal*, https://doi.org/10.1016/j.csbj.2021.06.022.

[56] Yacchirema, D., & Chura, A. (2021). SafeMobility: An IoT-based system for safer mobility using machine learning in the age of COVID-19. *Procedia Computer Science*, 184, 524–531.

[57] Zivkovic, M., Bacanin, N., Venkatachalam, K., Nayyar, A., Djordjevic, A., Strumberger, I., & Al-Turjman, F. (2021). COVID-19 cases prediction by using hybrid machine learning and beetle antennae search approach. *Sustainable Cities and Society*, 66, 102669, https://doi.org/10.1016/j.scs.2020.102669.

[58] Mojjada, R. K., Yadav, A., Prabhu, A. V., & Natarajan, Y. (2020). Machine learning models for COVID-19 future forecasting. *Materials Today: Proceedings*, https://doi.org/10.1016/j.matpr.2020.10.962.

[59] Burdick, H., Lam, C., Mataraso, S., Siefkas, A., Braden, G., Dellinger, R. P., & Das, R. (2020). Prediction of respiratory decompensation in COVID-19 patients using machine learning: The READY trial. *Computers in Biology and Medicine*, 124, 103949, https://doi.org/10.1016/j.compbiomed.2020.103949.

[60] Malki, Z., Atlam, E. S., Hassanien, A. E., Dagnew, G., Elhosseini, M. A., & Gad, I. (2020). Association between weather data and COVID-19 pandemic predicting mortality rate: Machine learning approaches. *Chaos, Solitons & Fractals*, 138, 110137, https://doi.org/10.1016/j.chaos.2020.110137.

[61] Dandekar, R., Rackauckas, C., & Barbastathis, G. (2020). A machine learning-aided global diagnostic and comparative tool to assess effect of quarantine control in COVID-19 spread. *Patterns*, 1(9), 100145, https://doi.org/10.1016/j.patter.2020.100145.

[62] Doanvo, A., Qian, X., Ramjee, D., Piontkivska, H., Desai, A., & Majumder, M. (2020). Machine learning maps research needs in COVID-19 literature. *Patterns*, 1(9), 100123, https://doi.org/10.1016/j.patter.2020.100123.

[63] Shaban, W. M., Rabie, A. H., Saleh, A. I., & Abo-Elsoud, M. A. (2021). Detecting COVID-19 patients based on fuzzy inference engine and deep neural network. *Applied Soft Computing*, 99, 106906, https://doi.org/10.1016/j.asoc.2020.106906.

[64] Zahra, S. R., Chishti, M. A., Baba, A. I., & Wu, F. (2021). Detecting COVID-19 chaos driven phishing/malicious URL attacks by a fuzzy logic and data mining based intelligence system. *Egyptian Informatics Journal*, https://doi.org/10.1016/j.eij.2021.12.003.

[65] Ahmed, H. I., Nasr, A. A., Abdel-Mageid, S. M., & Aslan, H. K. (2021). DADEM: Distributed attack detection model based on big data analytics for the enhancement of the security of internet of things (IoT). *International Journal of Ambient Computing and Intelligence (IJACI)*, 12(1), 114–139.

[66] Sholla, S., Mir, R. N., & Chishti, M. A. (2021). A fuzzy logic-based method for incorporating ethics in the internet of things. *International Journal of Ambient Computing and Intelligence (IJACI)*, 12(3), 98–122.

[67] Balusa, B. C., & Gorai, A. K. (2021). Development of fuzzy pattern recognition model for underground metal mining method selection. *International Journal of Ambient Computing and Intelligence (IJACI)*, 12(4), 64–78.

10 Fuzzy Geometric-Based Cost-Optimization Technique for Company

Neha Kumari, Arun Prasad Burnwal,
and Neha Keshri

CONTENTS

10.1 Introduction 177
10.2 Literature Review 178
10.3 Proposed Method 181
10.4 Performance Analysis 184
10.5 Conclusions 191
10.6 References 191

10.1 INTRODUCTION

In the current scenario, the total number of companies increases rapidly based on company strategy and competitors. Each company has its own strategy to meet the goals of the company [1–2]. In a decision-making situation, the strategy is two types: first is based on an increasing factor, the second on the decreasing factor. The first category indicates that if the factors are increased, then company profit also increases. The second category indicates that if the factors are decreased, then company profit increases. The combination of both helps predict and analyze the company's life cycle. In the proposed method, an intelligent technique is proposed based on the fusion of linear programming, geometric programming, and fuzzy logic. *Linear programming* is a numerical optimization technique which is used to optimize objective function based on related constraints depending on the situation. *Geometric programming* is a nonlinear programming which is used to efficiently optimize numerical problems based on related objective function and constraints in the posynomial environment instead of the polynomial environment. Sometimes, this is known as optimization purpose, which helps model the application based on different factors and information. It is also used to optimize nonlinear parameters efficiently as a nature-inspired optimization system [3]. The method used in this manuscript helps in several applications because it uses fuzzy logic to model the different parameters efficiently. Fuzzy logic is used to optimize imprecise information with the help of membership functions. Fuzzy logic is a type of soft computing methodology that produces soft results using an approximation method. Soft computing, in conjunction with wireless networks and wireless sensor networks, is utilized to tackle a variety of concerns and problems [4]. The suggested method's combination of the aforementioned strategies aids in the effective optimization of the proposed mathematical model based on restrictions.

The rest of the paper is organized, as the next section discusses several works based on optimization. The next section discusses the detailed illustration of the proposed method. The next section deals with simulation analysis and its formulation. And the last section discusses the conclusion of the paper.

10.2 LITERATURE REVIEW

In this section, some works have been discussed and organized for the purpose of optimization. There are several optimization techniques mentioned in this section, such as linear or nonlinear optimization, fuzzy optimization, and nature optimization. Some of the descriptions of these works are shown when S. K. Das [5] proposed a method of application design system and its management. It helps in several application management that deals with several challenges and issues. It helps model several information systems based on smart applications. It helps model several security management systems based on emergency management and application. It helps give new guidelines that help model several issues in terms of solution. This solution helps model several services based on real-life application management. It helps in several monitoring and application management based on emergency and security modeling based on services. Keerthika and Shanmugapriya [6] designed a method with a combination of passive and active attack for illustration. This illustration is based on some countermeasures that help in vulnerabilities system. It helps model the application based on environment analysis to help deal with a protection system based on commercial analysis. The communication of the system is based on the deployment of some challenges along with issues. It helps in defensive analysis based on vulnerability of some factors of information. Shaban et al. [7] designed a patient analysis system used for inference system and design. It is based on several information that help model several neural network systems. The work is based on a fuzzy inference modeling system that helps model some issues of COVID-19. The issue is based on deep learning and fuzzy system modeling. It helps model several network information based on strategy management. It helps validate cross analysis and validation to help in accuracy modeling. It helps model several detection systems for coronavirus analysis to help and model based on prevention system. Das et al. [8] designed an application management illustration for wireless network systems and services. It helps model several applications based on different security and challenge system management. It helps in issues of management based on several parts of the communication system. It helps adopt several solutions based on some higher analysis and management. The work is based on complexity management that helps model several solutions within the context of management based on different variations of the wireless network. Wan and Chen

[9] designed a strategy for energy analysis and mechanism for harvesting. The work is based on the WSN purpose of modeling. It helps model several cooperative analyses for node analysis. It defines some probability based on relay node detection. The main purpose of this analysis is to solve network performance based on certain factors. It helps model the application and save actual energy. It uses mathematical modeling for analyzing data and parameters. It helps enhance the energy based on solar energy systems and its cooperation. Zahra et al. [10] designed a method for a data-driven system that is based on intelligence system maintenance and information system. The work is based on URL system of information and modeling that help analyze and design several malicious and phishing information systems. The work is completely based on fuzzy logic modeling. It helps deal with several uncertainty management for handling pandemic analysis. It helps model several control and unprecedented information systems. The work helps model cybercriminal information analyses to deal with several ransomware information systems for impact analysis. De et al. [11] proposed an illustration for the purpose of challenges and application management of services. This service is based on the application of a wireless sensor network and its variations. It helps model several challenges and application management. It helps deal with several algorithms of wireless network based on variations and analysis. It helps guide several working principles and information systems based on service management. It helps deal with algorithm analysis that helps in managing several applications of the system. Misra et al. [12] designed an implementation method based on the fusion of FPGA and NLOS. The work is based on distance analysis and its estimation system. It helps model several applications that help elderly modeling. The work is designed for an indoor system that helps in WSN. It helps in location analysis based on the ZigBee network. The work uses a programmable gate array system and its modeling. This modeling uses artificial neural network to estimate different errors and improve network lifetime. It uses a hybridization method for modeling several complexities based on suitable analysis. Pahar et al. [13] proposed a method for cough analysis and its different types of classification based on COVID-19. The work is based on a machine learning algorithm and its different inherent elements. The work also uses some smartphone systems for resolving issues. The disease is based on two types; first is normal cough, and second is forced cough. It is based on some datasets used for COVID-19 patients. Das et al. [14] designed a book of architectural solution system based on a wireless network. This service is not only based on network but also on several systems and information based on the architecture of the network. It helps in modifications based on the architecture of the system. It helps model several issues of the network. Several issues are used and deal with the system. Some of the issues mentioned include energy efficiency system, network lifetime system, resource management, data aggregation system, etc. It helps model several solutions and in the security management of wireless network. Wang and Hu [15] designed a hole-detection method for handling several issues based on WSN. The network is based on a clustering method and algorithm that uses some gap-coverage analysis. It helps analyze multihop management systems for rational deployment. It helps in

distance parameter systems and vulnerability detection to help in coverage and its parameter modeling. It overcomes the limitation of several determination systems for edge node modeling. It helps determine random walk connection and its management. Kyriazos et al. [16] designed a method for COVID-19 quarantine system based on machine learning. It is based on several well-being score systems. The work is based on a machine learning system for modeling different datasets based on quarantine information available in websites. It is based on Diener's subjective well-being. It is based on factor analysis and management, where data is classified based on information, and it treated 25% as information. Finally, it helps produce several information based on the fusion of machine learning and exploratory graph analysis. Das et al. [17] proposed a book that helps model several applications based on industrial application and formulation. It helps model several information based on machine learning systems and modeling. The content of this book deal with several information based on a decision-making system and prediction analysis. It helps deal with some applications based on natural language processing, machine learning, computer vision, image processing, etc. Each application and system is based on several information and modeling for dealing with and analyzing information model. Temene et al. [18] illustrated a survey based on mobility analysis and prediction for WSN. The work is based on IoT and WSN both for detailed illustration. It helps model several mobile nodes. There are several mobile nodes that play different roles, such as the sink node, mobile node, source node, etc. The combination of all nodes helps model several congestions and its related mitigations. It helps in the predecessor analysis of IoT that helps in several directions. The work helps model several evaluations based on different algorithms. Kassania et al. [19] designed a method based on COVID-19 which is part of the automatic detection of the COVID disease. Two basic components are used in this article, namely, CT image and X-ray; based on these two medical information, disease is detected. A machine learning tool is used to detect the information. There are several symptoms used in this model, including fever, sore throat, and cough. There are several machine learning algorithms used in this article, such as ResNet, DenseNet, MobileNet, NASNet, etc., for predicting the disease efficiently and effectively. De et al. [20] designed a book for wireless sensor network. It helps model several applications based on services and management. It helps deal with several information systems and management for key area management. The work is based on nature-inspired applications and computing that help model several issues. It helps implement several applications and computation information systems. Information of this book is distributed in the form of a bio- and nature-inspired system. It helps model and design several applications for single-objective and multiobjective optimization systems. Yousefpoor et al. [21] designed a secure method for WSN as review paper that helps model several issues in the network. The work is based on a data aggregation method that helps reduce attack in the system. It helps in countermeasures for several issues with the context of attack measurement. This review is also based on industrial internet of things and its modeling. It helps in managing several issues within the context of solution measurement. It helps save energy and increase security of the system based on authentication. Alballa and Al-Turaiki [22] designed a method for severity risk analysis based on different predictions of severity along with mortality and the diagnosis of different diseases based on COVID-19. It is based on a machine learning technique for analysis of different parameters. The article is based on several types of data based on laboratory and clinical information. All information is based on analysis and prediction based on different features of prognosis and different types of implementations. Das et al. [23] designed a system and modeling for wireless network and applications. It helps model several information based on service management and application modeling. The work is based on several information and management systems, such as energy resource management and modeling. It helps deal with several security and privacy management systems that help in its design and enhancement [24–26]. The work is based on troubleshooting and automation system for network lifetime management and its enhancement. It gives several protocols for design perspective and modeling. Zhang and Mao [27] designed a multifactor system for authentic purposes. The work is based on a protocol system that helps model the application. It helps model several recognitions to exercise the physical system. It is based on the ZigBee network system that helps model several scope identifications. It helps in security analysis and recognition of several applications based on component analysis and management. It helps model several information based on radio frequency analysis. It helps design the system based on security analysis for connection management of the network. Ebinger et al. [28] designed a method based on algorithm analysis and prediction for COVID-19 patients and disease management. The prediction is observed during the time of hospitalization. Different types of variations are measured based on factors and systems of patients' length of stay in the hospital. It is also called LOS based on electronic length analysis of different factors of information. There are three models designed for the purpose of analysis, namely, hospital information based on days 1, 2, and 3. The model is analyzed based on some factors of information and its management.

10.3   PROPOSED METHOD

The entire suggested approach is outlined in this section based on numerous factors. The suggested approach combines linear programming, geometric programming, and fuzzy logic. The purpose of this fusion is to optimize cost of the company based on different variables of service and parameters. Equation 10.1 shows the linear model of the proposed method based on some constraints. The generated dataset is shown in Tables 10.1 to 10.3 for the fuzzy variables "low," "medium," and "high." Each of these fuzzy linguistic variables is formulated in the data range of 6,000 to 10,000 based on three decision variables, namely, $x1$, $x2$, and $x3$, for cost minimization of the objective function $Z1$. Based on the generated dataset, it has been found that cost increases based on "limit1" value. But cost is decreased based on the nature of fuzzy linguistic variable when it increases.

Minimize: $Z1 = x1 + x2 + x3;$
Subject to constraints: $p11x1 + p21x2 + p31x3 \geq limit1;$
                          $P12x1 + p22x2 + p32x3 \geq limit1;$ (10.1)
                          $p13x1 + p23x2 + p33x3 \geq limit1;$
                          $x1, x2, x3 > 0;$

Where x1, x2, and x3 are the controlling decision variables for three objectives, namely, "parameter 1," "parameter 2," and "parameter 3," and p1i, p2i, and p3i

TABLE 10.1
Dataset of Linear Model Under "Low" Fuzzy Linguistic Variable

Decision Variable	Limit1				
	6,000	7,000	8,000	9,000	10,000
x1	0.000000	26.92308	23.52941	0.00000	0.00000
x2	25.42373	16.15385	0.000000	26.41509	40.90905
x3	10.16949	0.000000	23.52941	24.52830	18.18182
Minimize Cost (Z1)	35.59322	43.07692	47.05882	50.94340	59.09091

TABLE 10.2
Dataset of Linear Model Under "Medium" Fuzzy Linguistic Variable

Decision Variable	Limit1				
	6,000	7,000	8,000	9,000	10,000
x1	0.000000	0.000000	6.557377	0.000000	0.000000
x2	21.42857	8.816121	0.000000	4.511278	0.000000
x3	0.000000	15.86902	19.67213	24.81203	35.71429
Minimize Cost (Z1)	21.42857	24.68514	26.22951	29.32331	35.71429

TABLE 10.3
Dataset of Linear Model Under "High" Fuzzy Linguistic Variable

Decision Variable	Limit1				
	6,000	7,000	8,000	9,000	10,000
x1	0.000000	0.000000	0.000000	0.00000	0.000000
x2	0.000000	0.000000	0.000000	0.00000	0.000000
x3	12.50000	14.58333	17.77778	20.0000	22.22222
Minimize Cost (Z1)	12.50000	14.58333	17.77778	20.0000	22.22222

are the controlling constraints for three different parameters, where i = 1, 2, 3 for different values of the fuzzy linguistic variables.

Equation 10.2 shows the geometric model of the proposed method based on some constraints. The generated dataset is shown in Tables 10.4 to 10.6 for the fuzzy variables "low," "medium," and "high." Each of these fuzzy linguistic variable is formulated in the data range 6,000 to 10,000 based on three decision variables, namely, $x1$, $x2$, and $x3$, for cost minimization of the objective function $Z2$. Based on the generated dataset, it has been found that cost is increased based

TABLE 10.4
Dataset of Geometric Model Under "Low" Fuzzy Linguistic Variable

	Limit2				
Decision Variable	6,000	7,000	8,000	9,000	10,000
x1	0.000000	0.000000	0.000000	0.000000	83.333333
x2	37.50000	0.000000	0.000000	0.000000	0.000000
x3	0.000000	53.84615	57.14286	69.23077	0.000000
Minimize Cost (Z2)	6.123724	7.337994	7.559289	8.320503	9.128709

TABLE 10.5
Dataset of Geometric Model Under "Medium" Fuzzy Linguistic Variable

	Limit2				
Decision Variable	6,000	7,000	8,000	9,000	10,000
x1	0.000000	0.000000	0.000000	0.000000	0.000000
x2	0.000000	0.000000	0.000000	0.000000	0.000000
x3	21.42857	2500000	27.58621	31.03448	35.71429
Minimize Cost (Z2)	4.629100	5.00000	5.252257	5.570860	5.976143

TABLE 10.6
Dataset of Geometric Model Under "High" Fuzzy Linguistic Variable

	Limit2				
Decision Variable	6,000	7,000	8,000	9,000	10,000
x1	0.000000	0.000000	0.000000	0.000000	0.000000
x2	0.000000	17.94872	0.000000	0.000000	0.000000
x3	12.5000	0.000000	17.77778	20.0000	22.22222
Minimize Cost (Z2)	3.535534	4.236593	4.216370	4.72136	4.714045

on "limit2" value. But cost is decreased based on the nature of fuzzy linguistic variable when it increases.

Minimize: $Z2 = (x1)^{0.5} + (x2)^{0.5} + (x3)^{0.5}$;
Subject to constraints: $p11x1 + p21x2 + p31x3 \geq limit2$;
                          $P12x1 + p22x2 + p32x3 \geq limit2$;   (10.2)
                          $p13x1 + p23x2 + p33x3 \geq limit2$;
                                    $x1, x2, x3 > 0$;

10.4 PERFORMANCE ANALYSIS

The specifics of simulation and analysis are illustrated in this section. The suggested technique is tested using the LINGO optimization program, which is based on a combination of linear and geometric formulations. Table 10.7 displays the fundamental simulation settings. Windows 11 is utilized in this simulation, along with MS Office 2016 and optimization applications. The suggested technique employs ten mathematical models which are both linear and nonlinear in character. The total linear model used is 5, and the total nonlinear model used is 5. The proposed method is evaluated in five iterations based on the number of nodes as 1,000, 2,000, 3,000, 4,000, and 5,000. So here the minimum sensor node is 1,000 and the maximum sensor node is 5,000. The total objective function used is 10 for the combination of linear and quadratic

TABLE 10.7
Details of Simulation Parameters

Parameter	Description
Windows	Windows 11
Linear model	1
Geometric model	1
MS Office	2016
Optimization software	LINGO
Total optimization models	2
Minimum limit value	6000
Maximum limit value	10000
Total objective functions	2
Nature of the objectives	Linear and nonlinear
Total constraints	6
Constraints for linear model	3×1
Constraints for geometric model	3×1
Total linguistic variables	3
Linguistic variables used	Low, medium, and high

programming. In linear programming, 5 objective functions are used with 3 × 5 constraints, and in quadratic programming, also 5 objectives are used with 3 × 5 constraints. The total linguistic variables used is three, namely, "low," "medium," and "high."

Figures 10.1 to 10.6 show the output of linear formulation, where Figures 1 and 2 show first and last range of "limit1" for "low" linguistic variable. Figures 10.3 and 4 show first and last range of "limit1" for "medium" linguistic variable. Figures 10.5 and 10.6 show first and last range of "limit1" for "high" linguistic variable. The figures indicate that the first range is lower than the second range based on "limit1" range.

Figures 10.7 to 10.12 show output of geometric formulation, where Figures 10.7 and 10.8 show first and last range of "limit2" for "low" linguistic variable. Figures 10.9 and 10.10 show first and last range of "limit2" for "medium" linguistic variable. Figures 10.11 and 10.12 show first and last range of "limit2" for "high" linguistic variable. The figures indicate that the first range is lower than the second range based on "limit2" range.

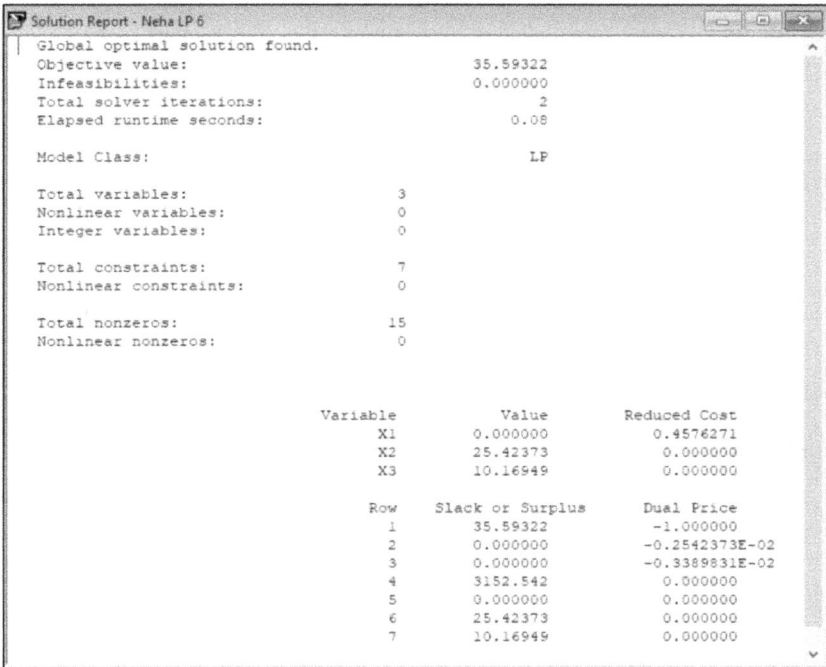

FIGURE 10.1 Linear formulation of cost minimization of first range based on "low" data.

```
Solution Report - Neha LP 10                                                    [_][□][x]
Global optimal solution found.
  Objective value:                                    59.09091
  Infeasibilities:                                     0.000000
  Total solver iterations:                                    2
  Elapsed runtime seconds:                                 0.08

  Model Class:                                               LP

  Total variables:                         3
  Nonlinear variables:                     0
  Integer variables:                       0

  Total constraints:                       7
  Nonlinear constraints:                   0

  Total nonzeros:                         15
  Nonlinear nonzeros:                      0


                            Variable            Value        Reduced Cost
                                  X1         0.000000      0.7272727E-01
                                  X2         40.90909           0.000000
                                  X3         18.18182           0.000000

                                 Row  Slack or Surplus          Dual Price
                                   1         59.09091           -1.000000
                                   2         0.000000      -0.2272727E-02
                                   3         0.000000      -0.3636364E-02
                                   4         3318.182           0.000000
                                   5         0.000000           0.000000
                                   6         40.90909           0.000000
                                   7         18.18182           0.000000
```

FIGURE 10.2   Linear formulation of cost minimization of last range based on "low" data.

```
Solution Report - Neha MP 6                                                     [_][□][x]
Global optimal solution found.
  Objective value:                                    21.42857
  Infeasibilities:                                     0.000000
  Total solver iterations:                                    1
  Elapsed runtime seconds:                                 0.06

  Model Class:                                               LP

  Total variables:                         3
  Nonlinear variables:                     0
  Integer variables:                       0

  Total constraints:                       7
  Nonlinear constraints:                   0

  Total nonzeros:                         15
  Nonlinear nonzeros:                      0


                            Variable            Value        Reduced Cost
                                  X1         0.000000           0.4642857
                                  X2         21.42857           0.000000
                                  X3         0.000000           0.000000

                                 Row  Slack or Surplus          Dual Price
                                   1         21.42857           -1.000000
                                   2         0.000000      -0.3571429E-02
                                   3         1714.286           0.000000
                                   4         2571.429           0.000000
                                   5         0.000000           0.000000
                                   6         21.42857           0.000000
                                   7         0.000000           0.000000
```

FIGURE 10.3   Linear formulation of cost minimization of first range based on "medium" data.

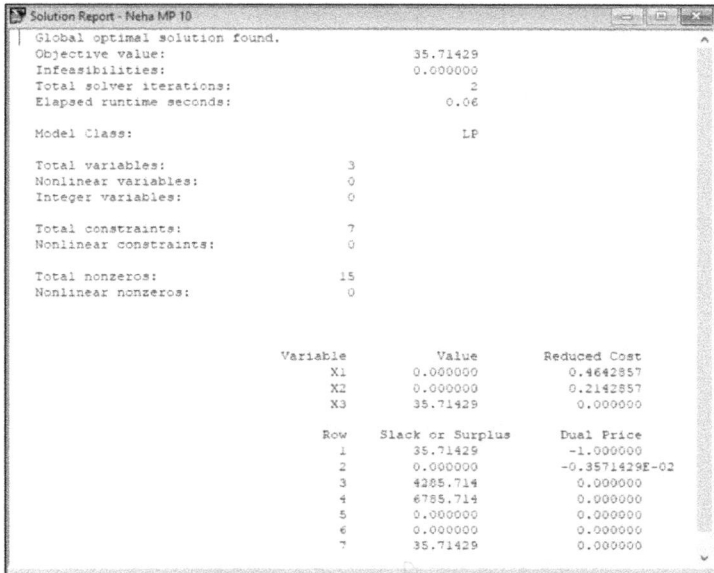

FIGURE 10.4 Linear formulation of cost minimization of last range based on "medium" data.

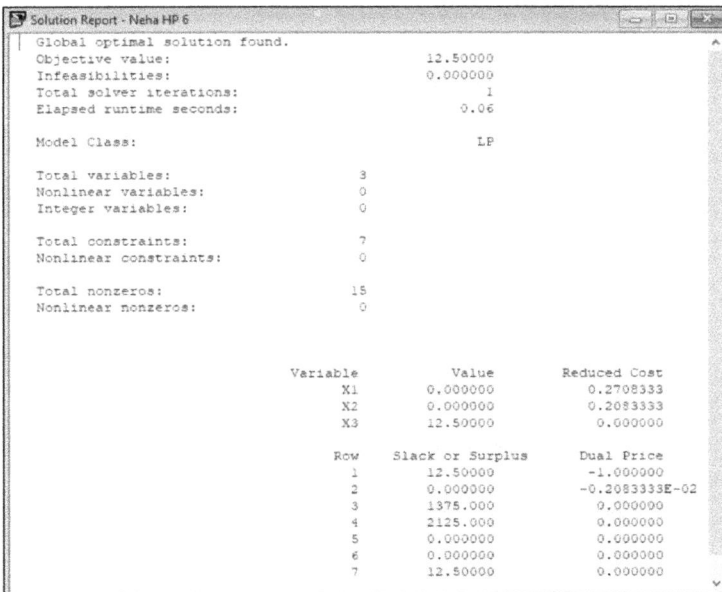

FIGURE 10.5 Linear formulation of cost minimization of first range based on "high" data.

```
Solution Report - Neha HP 10
Global optimal solution found.
Objective value:                              22.22222
Infeasibilities:                              0.000000
Total solver iterations:                             1
Elapsed runtime seconds:                          0.06

Model Class:                                        LP

Total variables:                    3
Nonlinear variables:                0
Integer variables:                  0

Total constraints:                  7
Nonlinear constraints:              0

Total nonzeros:                    15
Nonlinear nonzeros:                 0


                        Variable           Value        Reduced Cost
                              X1        0.000000           0.2222222
                              X2        0.000000           0.1555556
                              X3        22.22222            0.000000

                             Row    Slack or Surplus      Dual Price
                               1        22.22222           -1.000000
                               2        0.000000      -0.2222222E-02
                               3        3333.333            0.000000
                               4        5555.556            0.000000
                               5        0.000000            0.000000
                               6        0.000000            0.000000
                               7        22.22222            0.000000
```

FIGURE 10.6 Linear formulation of cost minimization of last range based on "high" data.

```
Solution Report - Neha GP 6
Local optimal solution found.
Objective value:                              6.123724
Infeasibilities:                              0.000000
Extended solver steps:                               5
Best multistart solution found at step:              1
Total solver iterations:                            25
Elapsed runtime seconds:                          0.26

Model Class:                                       NLP

Total variables:                    3
Nonlinear variables:                3
Integer variables:                  0

Total constraints:                  7
Nonlinear constraints:              1

Total nonzeros:                    15
Nonlinear nonzeros:                 3


                        Variable           Value        Reduced Cost
                              X1        0.000000      0.1414214E+08
                              X2        37.50000           0.000000
                              X3        0.000000      0.1414214E+08

                             Row    Slack or Surplus      Dual Price
                               1        6.123724           -1.000000
                               2        750.0000            0.000000
                               3        0.000000      -0.5103104E-03
                               4        3000.000            0.000000
                               5        0.000000            0.000000
                               6        37.50000            0.000000
                               7        0.000000            0.000000
```

FIGURE 10.7 Geometric formulation of cost minimization of first range based on "low" data.

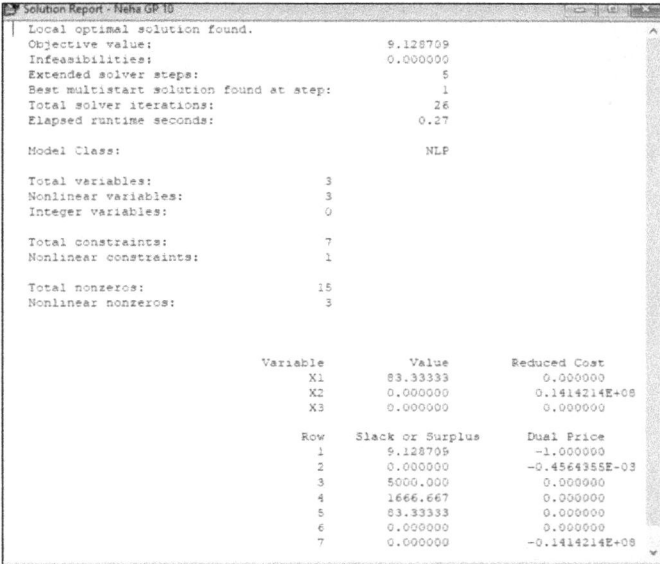

FIGURE 10.8 Geometric formulation of cost minimization of last range based on "low" data.

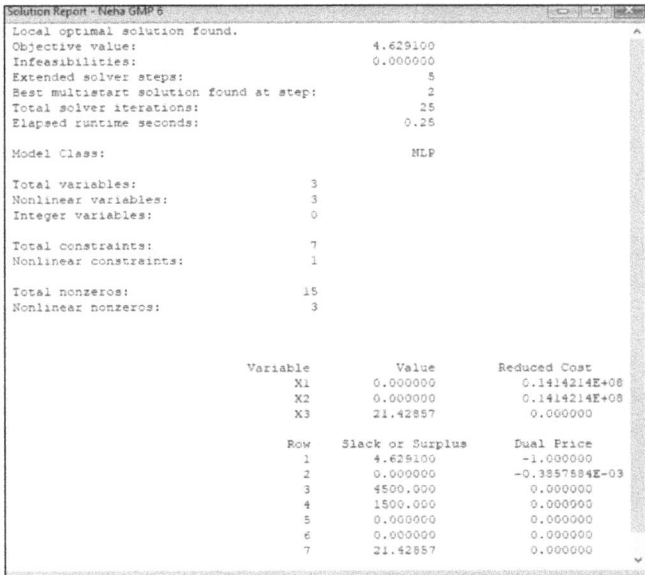

FIGURE 10.9 Geometric formulation of cost minimization of first range based on "medium" data.

```
Solution Report - Neha GMP 10
 Local optimal solution found.
 Objective value:                                5.976143
 Infeasibilities:                                0.000000
 Extended solver steps:                                 5
 Best multistart solution found at step:                1
 Total solver iterations:                              25
 Elapsed runtime seconds:                            0.27

 Model Class:                                         NLP

 Total variables:                    3
 Nonlinear variables:                3
 Integer variables:                  0

 Total constraints:                  7
 Nonlinear constraints:              1

 Total nonzeros:                    15
 Nonlinear nonzeros:                 3


                    Variable           Value        Reduced Cost
                          X1        0.000000      0.1414214E+08
                          X2        0.000000      0.1414214E+08
                          X3        35.71429           0.000000

                         Row  Slack or Surplus          Dual Price
                           1         5.976143           -1.000000
                           2         0.000000      -0.2988072E-03
                           3         4285.714            0.000000
                           4         6785.714            0.000000
                           5         0.000000            0.000000
                           6         0.000000            0.000000
                           7         35.71429            0.000000
```

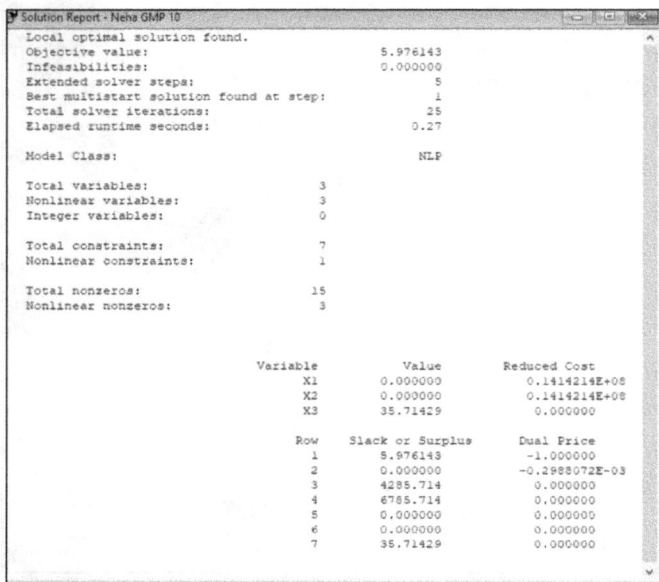

FIGURE 10.10 Geometric formulation of cost minimization of last range based on "medium" data.

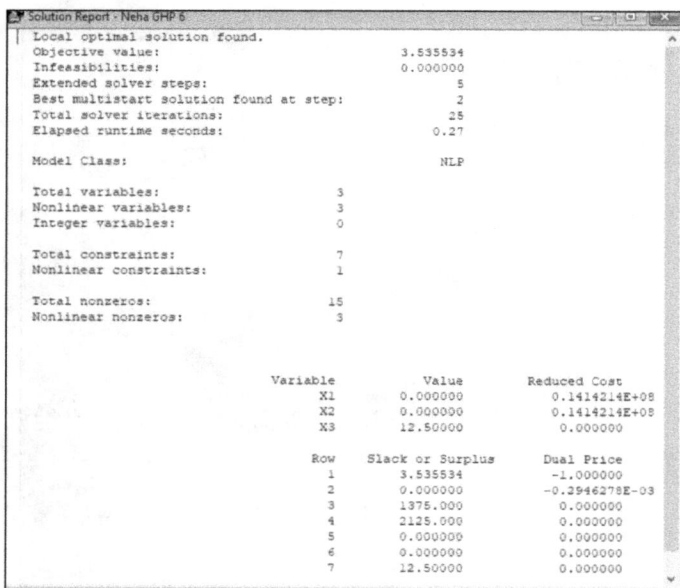

```
Solution Report - Neha GMP 6
 Local optimal solution found.
 Objective value:                                3.535534
 Infeasibilities:                                0.000000
 Extended solver steps:                                 5
 Best multistart solution found at step:                2
 Total solver iterations:                              25
 Elapsed runtime seconds:                            0.27

 Model Class:                                         NLP

 Total variables:                    3
 Nonlinear variables:                3
 Integer variables:                  0

 Total constraints:                  7
 Nonlinear constraints:              1

 Total nonzeros:                    15
 Nonlinear nonzeros:                 3


                    Variable           Value        Reduced Cost
                          X1        0.000000      0.1414214E+08
                          X2        0.000000      0.1414214E+08
                          X3        12.50000           0.000000

                         Row  Slack or Surplus          Dual Price
                           1         3.535534           -1.000000
                           2         0.000000      -0.2946278E-03
                           3         1375.000            0.000000
                           4         2125.000            0.000000
                           5         0.000000            0.000000
                           6         0.000000            0.000000
                           7         12.50000            0.000000
```

FIGURE 10.11 Geometric formulation of cost minimization of first range based on "high" data.

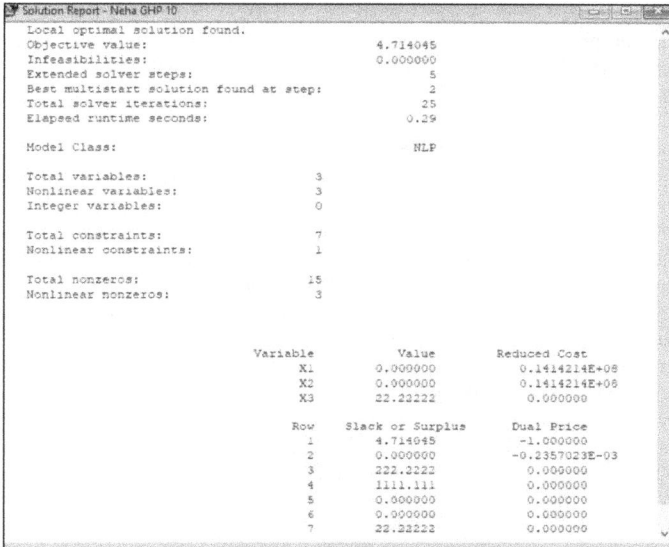

FIGURE 10.12 Geometric formulation of cost minimization of last range based on "high" data.

10.5 CONCLUSIONS

The suggested technique evaluates the combination of linear and geometric formulations for cost analysis. The suggested method's goal is to optimize the system's cost using linear and nonlinear formulations. *Geometric formulation* is another name for *nonlinear formulation*. In this system, the model is assessed into two limit values that show the model's maximum limit. Linear formulation is evaluated as "limit1," and geometric formulation is evaluated as "limit2," for the purpose of analysis and design. Both limit values are based on range 6,000 to 10,000. It helps evaluate several information based on objective function and its related constraints. It aids in the efficient optimization of nonlinear parameters by modeling numerous information-efficient parameters. Nonlinear parameters in this model are based on a number of constraints, which aid in the modeling of a variety of data. It helps display actual minimized cost based on related objective function and its related constraints. This optimization is done with the help of fuzzy logic to optimize several fuzzy linguistic variables.

10.6 REFERENCES

[1] Guo, W. Y., Guo, Y. Z., Zhong, M., & Yang, Z. Z. (2021). Optimizing the input amount to build cruise companies' direct selling channel based on consumers' choice behavior. *Research in Transportation Business & Management*, 100624, https://doi.org/10.1016/j.rtbm.2021.100624.

[2] Freije, I., de la Calle, A., & Ugarte, J. V. (2021). Role of supply chain integration in the product innovation capability of servitized manufacturing companies. *Technovation*, 102216, https://doi.org/10.1016/j.technovation.2020.102216.

[3] Roch-Dupre, D., Gonsalves, T., Cucala, A. P., Pecharroman, R. R., López-López, Á. J., & Fernandez-Cardador, A. (2021). Multi-stage optimization of the installation of Energy Storage Systems in railway electrical infrastructures with nature-inspired optimization algorithms. *Engineering Applications of Artificial Intelligence*, 104, 104370, https://doi.org/10.1016/j.engappai.2021.104370.

[4] Sharma, T., Mohapatra, A. K., & Tomar, G. S. (2021). A review of soft computing based cluster-heads selection algorithms in wireless sensor network. *Materials Today: Proceedings*, https://doi.org/10.1016/j.matpr.2021.02.724.

[5] Das, S. K. (2021). Smart design and its applications: Challenges and techniques. *Nature-Inspired Computing for Smart Application Design*, 1.

[6] Keerthika, M., & Shanmugapriya, D. (2021). Wireless sensor networks: Active and passive attacks-vulnerabilities and countermeasures. *Global Transitions Proceedings*, 2(2), 362–367.

[7] Shaban, W. M., Rabie, A. H., Saleh, A. I., & Abo-Elsoud, M. A. (2021). Detecting COVID-19 patients based on fuzzy inference engine and deep neural network. *Applied Soft Computing*, 99, 106906, https://doi.org/10.1016/j.asoc.2020.106906.

[8] Das, S. K., Maheswari, V., & Sharma, A. (2021). Wireless networks: Applications, challenges, and security issues. In *Architectural Wireless Networks Solutions and Security Issues* (pp. 1–10). Springer, Singapore.

[9] Wan, J., & Chen, J. (2022). AHP based relay selection strategy for energy harvesting wireless sensor networks. *Future Generation Computer Systems*, 128, 36–44.

[10] Zahra, S. R., Chishti, M. A., Baba, A. I., & Wu, F. (2021). Detecting COVID-19 chaos driven phishing/malicious URL attacks by a fuzzy logic and data mining based intelligence system. *Egyptian Informatics Journal*, https://doi.org/10.1016/j.eij.2021.12.003.

[11] De, D., Mukherjee, A., Das, S. K., & Dey, N. (2020). Wireless sensor network: Applications, challenges, and algorithms. In *Nature Inspired Computing for Wireless Sensor Networks* (pp. 1–18). Springer, Singapore.

[12] Misra, Y., Krishnaveni, K., & Rajasekaran, A. S. (2022). Implementation of NLOS based FPGA for distance estimation of elderly using indoor wireless sensor networks. *Materials Today: Proceedings*, https://doi.org/10.1016/j.matpr.2022.01.087.

[13] Pahar, M., Klopper, M., Warren, R., & Niesler, T. (2021). COVID-19 cough classification using machine learning and global smartphone recordings. *Computers in Biology and Medicine*, 104572, https://doi.org/10.1016/j.compbiomed.2021.104572.

[14] Das, S. K., Samanta, S., Dey, N., Patel, B. S., & Hassanien, A. E. (Eds.). (2021). *Architectural Wireless Networks Solutions and Security Issues*. Springer, Singapore.

[15] Wang, F., & Hu, H. (2021). Coverage hole detection method of wireless sensor network based on clustering algorithm. *Measurement*, 179, 109449, https://doi.org/10.1016/j.measurement.2021.109449.

[16] Kyriazos, T., Galanakis, M., Karakasidou, E., & Stalikas, A. (2021). Early COVID-19 quarantine: A machine learning approach to model what differentiated the top 25% well-being scorers. *Personality and Individual Differences*, 181, 110980, https://doi.org/10.1016/j.paid.2021.110980.

[17] Das, S. K., Das, S. P., Dey, N., & Hassanien, A. E. (Eds.). (2021). *Machine Learning Algorithms for Industrial Applications*. Springer, Switzerland.

[18] Temene, N., Sergiou, C., Georgiou, C., & Vassiliou, V. (2022). A survey on mobility in Wireless Sensor Networks. *Ad Hoc Networks*, 125, 102726, https://doi.org/10.1016/j.adhoc.2021.102726.

[19] Kassania, S. H., Kassanib, P. H., Wesolowskic, M. J., Schneidera, K. A., & Detersa, R. (2021). Automatic detection of coronavirus disease (COVID-19) in X-ray and CT images: A machine learning based approach. *Biocybernetics and Biomedical Engineering*, 41(3), 867–879.

[20] De, D., Mukherjee, A., Das, S. K., & Dey, N. (Eds.). (2020). *Nature Inspired Computing for Wireless Sensor Networks*. Springer, Singapore.

[21] Yousefpoor, M. S., Yousefpoor, E., Barati, H., Barati, A., Movaghar, A., & Hosseinzadeh, M. (2021). Secure data aggregation methods and countermeasures against various attacks in wireless sensor networks: A comprehensive review. *Journal of Network and Computer Applications*, 103118, https://doi.org/10.1016/j.jnca.2021.103118.

[22] Alballa, N., & Al-Turaiki, I. (2021). Machine learning approaches in COVID-19 diagnosis, mortality, and severity risk prediction: A review. *Informatics in Medicine Unlocked*, 100564, https://doi.org/10.1016/j.imu.2021.100564.

[23] Das, S. K., Samanta, S., Dey, N., & Kumar, R. (Eds.). (2020). *Design Frameworks for Wireless Networks*. Springer, Singapore.

[24] Ahmed, H. I., Nasr, A. A., Abdel-Mageid, S. M., & Aslan, H. K. (2021). DADEM: Distributed attack detection model based on big data analytics for the enhancement of the security of internet of things (IoT). *International Journal of Ambient Computing and Intelligence (IJACI)*, 12(1), 114–139.

[25] Sholla, S., Mir, R. N., & Chishti, M. A. (2021). A fuzzy logic-based method for incorporating ethics in the internet of things. *International Journal of Ambient Computing and Intelligence (IJACI)*, 12(3), 98–122.

[26] Balusa, B. C., & Gorai, A. K. (2021). Development of fuzzy pattern recognition model for underground metal mining method selection. *International Journal of Ambient Computing and Intelligence (IJACI)*, 12(4), 64–78.

[27] Zhang, J., & Mao, H. (2021). Multi-factor identity authentication protocol and indoor physical exercise identity recognition in wireless sensor network. *Environmental Technology & Innovation*, 101671, https://doi.org/10.1016/j.eti.2021.101671.

[28] Ebinger, J., Wells, M., Ouyang, D., Davis, T., Kaufman, N., Cheng, S., & Chugh, S. (2021). A machine learning algorithm predicts duration of hospitalization in COVID-19 patients. *Intelligence-based Medicine*, 100035, https://doi.org/10.1016/j.ibmed.2021.100035.

Index

A

ad hoc on-demand distance vector routing
(AODV), 24
ANOVA, 165
ant colony optimization (ACO), 4, 10, 17, 76
artificial intelligence, 9, 77–81, 117–130, 164, 167

B

base station (BS), 4, 10, 23, 75, 93, 137

C

CHs, 82–83
COVID-19, 4–9, 45, 63, 76–78, 80–81, 117–130,
155–170, 178–181

D

decision tree, 6, 124, 157, 159–161
deep learning, 45, 63, 76, 121, 123–128, 158,
160–162, 178
deep neural network, 79, 119

F

fuzzy logic, 6, 17, 30, 42, 45, 63–66, 77, 86, 93,
123, 142, 161, 177, 179, 181

G

game theory, 59, 63–64, 67

L

LINGO, 48, 103, 148, 184

M

machine learning, 4–9, 27, 44, 61, 77, 79–80,
117, 121, 123–128, 155–170, 179–180
mobile ad hoc network (MANET), 23, 27,
41–42, 44, 95, 97, 141
mobility, 26, 41, 61, 64–69, 96, 139, 163, 180

N

node, 4, 6–19, 24–30, 35–36, 41–52, 57–62,
64–66, 69–72, 75, 80–88, 93–111, 137–141,
144–152, 163, 179–180, 184
nonlinear formulation, 104, 109,
148, 191

O

OMNET++, 18, 76, 86

P

PSO, 6–12, 15–16, 18–19, 76, 81, 83–89

Q

quadratic programming, 41, 45, 48–49, 100,
104, 109, 111, 138, 142–144, 148, 185

R

RN, 83, 84

S

swarm optimization, 4, 9–10, 17, 83

U

utility function, 63

V

vehicular ad hoc network (VANET), 23, 27,
42, 95

W

wireless ad hoc network (WANET), 23–24, 138
wireless sensor network (WSN), 4, 5, 7, 9,
18, 24–27, 44–45, 57–58, 60–62, 76–78,
80, 86, 93, 96–97, 117, 138–141, 156, 157,
178–180